THE Lying Stones

Marrakech

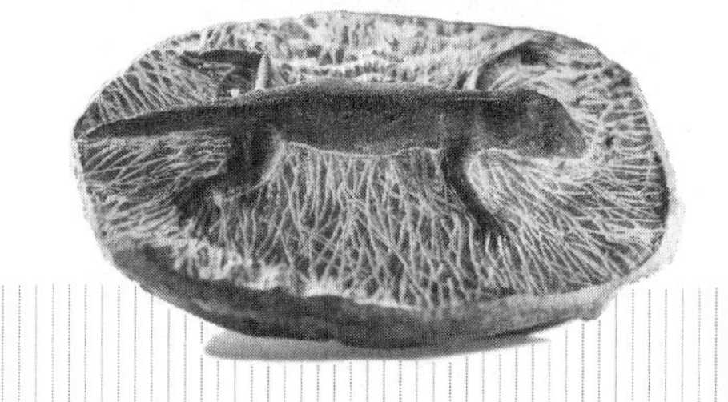

By the same author

Ontogeny and Phylogeny

Ever Since Darwin

The Panda's Thumb

The Mismeasure of Man

Hen's Teeth and Horse's Toes

The Flamingo's Smile

An Urchin in the Storm

Time's Arrow, Time's Cycle

Illuminations (with R. W. Purcell)

Wonderful Life

Bully for Brontosaurus

Finders, Keepers (with R. W. Purcell)

Eight Little Piggies

Dinosaur in a Haystack

Full House

Questioning the Millennium

Leonardo's Mountain of Clams
and the Diet of Worms

Rocks of Ages

Crossing Over (with R. W. Purcell)

THE Lying Stones

of Marrakech

Penultimate Reflections IN *Natural History*

Stephen Jay Gould

THREE RIVERS PRESS • NEW YORK

Published by Three Rivers Press, New York, New York.
Member of the Crown Publishing Group.

Random House, Inc. New York, Toronto, London, Sydney, Auckland
www.randomhouse.com

All of the essays contained in this work were previously published
by *Natural History* magazine.

Originally published in hardcover in 2000 by Harmony Books.

Printed in the United States of America

DESIGN BY LYNNE AMFT

Library of Congress Cataloging-in-Publication Data
Gould, Stephen Jay.
The lying stones of Marrakech: penultimate reflections in natural history /
by Stephen Jay Gould.
1. Natural history—Popular works. 2. Evolution—Popular works.
I. Title.
QH45.5.G74 2000
508—dc21 99-36148
CIP

ISBN 0-609-80755-2

10 9 8 7 6 5 4 3 2 1

First Paperback Edition

For Jack Sepkoski (1948–1999),
who brought me one of the greatest possible joys
a teacher can ever earn or experience:
to be surpassed by his students.
Offspring should not predecease their parents,
and students should outlive their teachers.
The times may be out of joint,
but Jack was born to set the order of life's history right—
and he did!

Contents

THE Lying Stones

Marrakech

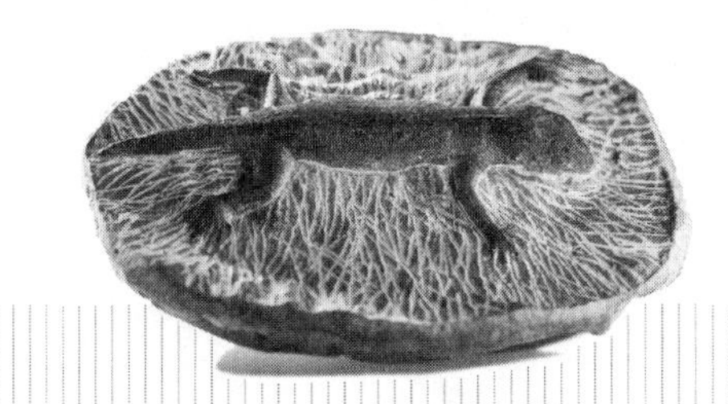

PREFACE

In the fall of 1973, I received a call from Alan Ternes, editor of *Natural History* magazine. He asked me if I would like to write columns on a monthly basis, and he told me that folks actually get paid for such activities. (Until that day, I had published only in technical journals.) The idea intrigued me, and I said that I'd try three or four. Now, 290 monthly essays later (with never a deadline missed), I look only a little way forward to the last item of this extended series—to be written, as number 300 exactly, for the millennial issue of January 2001. One really should follow the honorable principle of quitting while still ahead, a rare form of dignity chosen by such admirable men as Michael Jordan and Joe DiMaggio, my personal hero and mentor from childhood. (Joe died, as I put this book together, full of years and in maximal style and grace, after setting one last record—for number of times in receiving last rites and then rallying.) Our millennial transition may represent an arbitrary imposition of human decisions upon nature's true cycles, but what grander symbol for calling a halt and moving on could possibly cross the path of a man's lifetime? This ninth volume of essays will therefore be the penultimate book in a series that shall close by honoring the same decimal preference lying behind our millennial transition.

If this series has finally found a distinctive voice, I have learned this mode of speech in the most gradual, accumulating, and largely unconscious manner—against my deepest personal beliefs in punctuational change and the uniquely directive power (despite an entirely accidental origin) of human reason in evolution. I suppose I had read a bit of Montaigne in English 101, and I surely could spell the word, but I had no inkling about the definitions and traditions of the essay as a literary genre when Alan Ternes called me cold on that fine autumn day.

I began the series with quite conventional notions about writing science for general consumption. I believed, as almost all scientists do (by passively imbibing a professional ethos, not by active thought or decision), that nature speaks directly to unprejudiced observers, and that accessible writing for nonscientists therefore required clarity, suppression of professional jargon, and an ability to convey the excitement of fascinating facts and interesting theories. If I supposed that I might bring something distinctive to previous efforts in this vein, I managed to formulate only two vague personal precepts: first, I would try to portray all subjects at the same conceptual depth that I would utilize in professional articles (that is, no dumbing down of ideas to accompany necessary clarification of language); second, I would use my humanistic and historical interests as a "user friendly" bridge to bring readers into the accessible world of science.

Over the years, however, this mere device (the humanistic "bridge") became an explicit centrality, a feature that I permitted myself to accept (and regard as a source of comfort and pride rather than an idiosyncrasy to downplay or even to hide) only when I finally realized that I had been writing *essays,* not mere columns, all along—and that nearly five hundred years of tradition had established and validated (indeed, had explicitly defined) the essay as a genre dedicated to personal musing and experience, used as a gracious entrée, or at least an intriguing hook, for discussion of general and universal issues. (Scientists are subtly trained to define the personal as a maximally dangerous snare of subjectivity and therefore to eschew the first person singular in favor of the passive voice in all technical writing. Some scientific editors will automatically blue-pencil the dreaded *I* at every raising of its ugly head. Therefore, "popular science writing" and "the literary essay" rank as an ultimately disparate, if not hostile, pairing of immiscible oil and water in our usual view—a convention that I now dream about fracturing as a preeminent goal for my literary *and* scientific life.)

I have tried, as these essays developed over the years, to expand my humanistic "take" upon science from a simple practical device (my original intention, insofar as I had any initial plan at all) into a genuine emulsifier that might fuse the literary essay and the popular scientific article into something distinctive, something that might transcend our parochial disciplinary divisions for the benefit of both domains (science, because honorable personal expression by competent writers can't ever hurt; and composition, because the thrill of nature's factuality should not be excluded from the realm of our literary efforts). At the very least, such an undertaking can augment the dimensionality of popular scientific articles—for we lose nothing of science's factual beauty and meaning, while we add the complexity of how we come to know (or fail to learn) to conventional accounts of what we think we know.

As this series developed, I experimented with many styles for adding this humanistic component about how we learned (or erred) to standard tales about what, in our best judgment, exists "out there" in the natural world—often only to demonstrate the indivisibility of these two accounts, and the necessary embeddedness of "objective" knowledge within worldviews shaped by social norms and psychological hopes. But so often, as both Dorothy and T. S. Eliot recognized in their different ways, traditional paths may work best and lead home (because they have truly withstood the test of time and have therefore been honed to our deep needs and best modes of learning, not because we fall under their sway for reasons of laziness or suppression).

Despite conscious efforts at avoidance, I find myself constantly drawn to biography—for absolutely nothing can match the richness and fascination of a person's life, in its wondrous mixture of pure gossip, miniaturized and personalized social history, psychological dynamics, and the development of central ideas that motivate careers and eventually move mountains. And try as I may to ground biography in various central themes, nothing can really substitute for the sweep and storytelling power of chronology. (I regard the Picasso Museum in Paris and the Turner Wing of the Tate Gallery in London as my two favorite art museums because each displays the work of a great creator in the strict chronological order of his life. I can then devise whatever alternative arrangement strikes my own fancy and sense of utility—but the arrow of time cannot be replaced or set aside; even our claims for invariance must seek constant features of style or subject *through* time's passage.)

So I have struggled, harder and more explicitly than for anything else in my life as a writer, to develop a distinctive and personal form of essay to treat great scientific issues in the context of biography—and to do so not by the factual chronology of a life's sorrows and accomplishments (a noble task requiring the amplitude of a full book), but rather by the intellectual synergy between a person and the controlling idea of his life. In this manner, when the conceit works, I can capture the essence of a scientist's greatest labor, including the major impediments and insights met and gathered along the way, while also laying bare (in the spare epitome demanded by strictures of the essay as a literary form of limited length) the heart of a key intellectual concept in the most interesting microcosm of a person's formulation and defense.

The first three parts of this book apply this strategy to three different times, places, subjects, and worldviews—an extended test of my claim for a distinctive voice based on applying biographical perspectives to the illumination of key scientific concepts and their history (following the basic strategy, in each essay, of linking a person's central operating idea, the focus of a professional life in development, to an important concept in human understanding of the natural world—in other words, to summarize the range and power of a principle by exemplifying its role in the intellectual development of a particularly interesting scientist). Thus I have tried to encapsulate, in the unforgiving form of an essay, the essence of both a person (as expressed in the controlling idea of his scientific life) and a concept (through the quintessentially human device of displaying its development in an individual life).

Part I treats the most fascinating period in my own subject of paleontology, the premodern struggle (sixteenth to early eighteenth centuries) to understand the origin of fossils while nascent science struggled with the deepest of all questions about the nature of both causality and reality themselves. Are fossils the remains of ancient organisms on an old earth, or manifestations of a stable and universal order, symbolically expressed by correspondences among nature's three kingdoms of animal, vegetable, and mineral, with fossils arising entirely within the mineral kingdom as analogs of living forms in the other two realms? No subject could be more crucial, and no alternative view more eerily unfamiliar, than this particular battleground for the nature of reality. I present three variations upon this theme, each biographically expressed: the early-eighteenth-century tale of paleontology's most famous hoax, combined with a weirdly similar story from modern Morocco; the linkage of the unknown Stelluti to the preeminent Galileo through their friendship, and through a common error that unites the master's original view of Saturn with Stelluti's erroneous belief that petrified wood arose in the mineral kingdom; and finally, a "reversed" biography expressed in terms of an organism under study (the brachiopod fossils that were once called "vulva stones" for their resemblance to female genitalia) rather than a person pursuing the investigation.

Part II then discusses the greatest conjunction of a time, a subject, and a group of amazing people in the history of natural history: late-eighteenth- and early-nineteenth-century France, when a group including some of the most remarkable intellects of the millennium invented the scientific study of natural history in an age of revolution. Georges Buffon establishes a discipline, by the grandest route of virtually defining a new and historically based way of knowing, in the forty-four volumes of his eminently literary *Histoire naturelle,* and then loses public recognition, for interesting and understandable reasons, in the midst of his ubiquity. Antoine Lavoisier, the most stunningly incisive intellect I have ever encountered, literally adds a new dimension to our understanding of nature in the geometry of geological mapping, his one foray (amidst intentions cut short by the guillotine) into my profession. Jean-Baptiste Lamarck belies his own unfairly imposed reputation for error and inflexibility with a heartrending reassessment of the foundations of his own deepest belief—in an odyssey that begins with a handwritten comment and drawing, inked by Lamarck into his own copy of his first evolutionary treatise, and here discovered and presented for the first time.

Part III then illustrates the greatest British challenge to this continental pre-eminence: the remarkable, and wonderfully literate, leading lights of Victorian science in Darwin's age of turmoil and reassessment: the heart of Lyell's uniformitarianism as seen (literally) by visiting the site of his most famous visual image, the pillars of Pozzuoli, used as a frontispiece to all editions of his *Principles of Geology;* Darwin's own intellectual development from such an unpromising temperament and early training to an ultimately understandable role as the most gentle, yet thorough revolutionary in the history of science; Richard Owen's invention of dinosaurs as an explicit device to subvert the evolutionary views of a generation before Darwin; and Alfred Russel Wallace on Victorian certainties and subsequent unpredictabilities.

The last three parts of this book do not invoke biography so explicitly, but they also use the same device of embodying an abstraction within a particular that can be addressed in sufficient detail and immediate focus to fit within an essay. The interlude of part IV presents some experiments in the different literary form of short takes (op-ed pieces, obituary notices, and even, in one case, an introductory statement for Penguin CD's series of famous classical compositions). Here I include six attempts (the literal meaning of *essay)* to capture the most elusive and important subject of all: the nature and meaning of excellence, expressed as a general statement about substrates (chapter 11) followed by five iterations on the greatness of individuals and their central passions across a full range of human activity—for excellence must be construed as a goal for all varieties of deeds and seasons, not only for mental categories—from bodily grace and dignity within domains debased by the confusion of celebrity with stature; to distinctive individuality within corporate blandness; to the intellectual innovations more commonly cited by scholars to exemplify this most precious (and uncommon) of human attributes.

Part V, on scientific subjects with more obvious and explicit social consequences (and often, unacknowledged social origins as well), also uses biography, but in a different way to link past stories with present realities—to convey the lesson that claims for objectivity based on pure discovery often replay episodes buried in history, and proving (upon exhumation and linkage) that our modern certainties flounder within the same complexities of social context and mental blockage: Spencer's social Darwinism, the Triangle Shirtwaist fire, and modern eugenics (chapter 17); contemporary boasts about the discovery of genes for specific behaviors, Davenport's heritability of wanderlust, and the old medical theory of humors (18); Dolly the cloned sheep, the nature of identical twins, and the decapitation of Louis XVI (19); J. B. S. Haldane on the "humaneness" of poison gas in warfare, and the role and status of unpredictability in science (20).

Finally, part VI abandons biography for another device of essayists: major themes (about evolution's different expression across scales of size and time) cast into the epitome of odd or intriguing particulars: fossil embryos nearly 600 million years old (21); three stories about measurable evolution in snails, lizards, and fishes (22), conventionally misinterpreted as modest enough to prove the efficacy of Darwin's mechanism extended across the immensity of geological time, but far too rapid and convulsive to convey any such meaning when properly read at this grand and unfamiliar scale; and avoidance in antipathy among several Christian groups (23) that "share" Jerusalem's Church of the Holy Sepulchre (the traditional site of Christ's crucifixion).

At this equipoise, with one more foray into the breach yet to come, I can only thank readers who have joined me on this rocky journey. For only the conjunction of growing fellowship and increasing knowledge—a loop of ethical and intellectual, emotional and rational feedback that positively rings with the optimism of potential survival, maybe even transcendence, in this endlessly fascinating world of woe—can validate the accident of our existence by our free decision to make maximal use of those simple gifts that nature and evolution have granted us.

I

Episodes in the Birth of Paleontology

The Nature of Fossils and the History of the Earth

1

The Lying Stones of Marrakech

WE TEND TO THINK OF FAKERY AS AN ACTIVITY DEDICATED to minor moments of forgivable fun (from the whoopie cushion to the squirting lapel flower), or harmless embellishment (from my grandfather's vivid eyewitness tales of the Dempsey-Firpo fight he never attended, to the 250,000 people who swear they were there when Bobby Thomson hit his home run in a stadium with a maximal capacity of some fifty thousand).

But fakery can also become a serious and truly tragic business, warping (or even destroying) the lives of thousands, and misdirecting entire professions into sterility for generations. Scoundrels may find the matrix of temptation irresistible, for immediate gains in money and power can be so great, while human gullibility grants the skillful forger an apparently limitless field of operation. The Van Gogh *Sunflowers,* bought in 1987 by a Japanese insurance company for nearly 25 million pounds sterling—then a record price for a painting—may well be a forged copy made around 1900 by the stockbroker and artist manqué Emile Schuffenecker. The phony Piltdown Man, artlessly confected from the jaw of an orangutan and a modern human cranium, derailed the profession of paleoanthropology for forty years, until exposed as a fake in the early 1950s.

Earlier examples cast an even longer and broader net of disappointment. A large body of medieval and Renaissance scholarship depended upon the documents of Hermes Trismegistus (Thrice-Great Hermes), a body of work attributed to Thoth, the Egyptian god of wisdom, and once viewed as equal in insight (not to mention antiquity) to biblical and classical sources—until exposed as a set of forgeries compiled largely in the third century A.D. And how can we possibly measure the pain of so many thousands of pious Jews, who abandoned their possessions and towns to follow the false messiah Shabbetai Tzevi to Jerusalem in the apocalyptic year of 1666—only to learn that their leader, imprisoned by the sultan and threatened with torture, had converted to Islam, been renamed Mehmed Efendi, and made the sultan's personal doorkeeper.

The most famous story of fraud in my own field of paleontology may not qualify for this first rank in the genre, but has surely won both general fame and staying power by persistence for more than 250 years. Like all great legends, this story has a canonical form, replete with conventional moral messages, and told without any variation in content across the centuries. Moreover, this standard form bears little relationship to the actual course of events as best reconstructed from available evidence. Finally, to cite the third common property of such legends, a correction of the conventional tale wins added and general value in teaching us important lessons about how we use and abuse our own history. Thus, the old story merits yet another retelling—which I first provide in the canonical (and false) version known to so many generations of students (and no doubt remembered by many readers from their college courses in natural science).

In 1726, Dr. Johann Bartholomew Adam Beringer, an insufferably pompous and dilettantish professor and physician from the town of Würzburg, published a volume, the *Lithographiae Wirceburgensis* (Würzburg lithography), documenting in copious words and twenty-one plates a remarkable series of fossils that he had found on a mountain adjacent to the city. These fossils portrayed a large array of objects, all neatly exposed in three-dimensional relief on the surface of flattened stones. The great majority depicted organisms, nearly all complete, including remarkable features of behavior and soft anatomy that had never been noted in conventional fossils—lizards in their skin, birds complete with beaks and eyes, spiders with their webs, bees feeding on flowers, snails next to their eggs, and frogs copulating. But others showed heavenly objects—comets with tails, the crescent moon with rays, and the sun all effulgent with a glowing central face of human form. Still others depicted Hebrew letters, nearly all spelling out the tetragrammaton, the ineffable name of God—YHVH, usually transliterated by Christian Europe as "Jehovah."

Beringer did recognize the difference between his stones and conventional fossils, and he didn't state a dogmatic opinion about their nature. But he didn't doubt their authenticity either, and he did dismiss claims that they had been carved by human hands, either recently in an attempt to defraud, or long ago for pagan purposes. Alas, after publishing his book and trumpeting the contents, Beringer realized that he had indeed been duped, presumably by his students playing a prank. (Some sources say that he finally acknowledged the trickery when he found his own name written in Hebrew letters on one stone.) According to legend, the brokenhearted Beringer then impoverished himself by attempting to buy back all copies of his book—and died dispirited just a few years later. Beringer's false fossils have been known ever since as *Lügensteine,* or "lying stones."

To illustrate the pedigree of the canonical tale, I cite the version given in the most famous paleontological treatise of the early nineteenth century, Dr. James Parkinson's *Organic Remains of a Former World* (volume 1, 1804). Parkinson, a physician by training and a fine paleontologist by avocation, identified and gave his name to the degenerative disease that continues to puzzle and trouble us today. He wrote of his colleague Beringer:

> One work, published in 1726, deserves to be particularly noticed; since it plainly demonstrates, that learning may not be sufficient to prevent an unsuspecting man, from becoming the dupe of excessive credulity. It is worthy of being mentioned on another account: the quantity of censure and ridicule, to which its author was exposed, served, not only to render his cotemporaries [*sic*] less liable to imposition; but also more cautious in indulging in unsupported hypotheses. . . . We are here presented with the representation of stones said to bear petrifactions of birds; some with spread, others with closed, wings: bees and wasps, both resting in their curiously constructed cells, and in the act of sipping honey from expanded flowers . . . and, to complete the absurdity, petrifactions representing the sun, moon, stars, and comets: with many others too monstrous and ridiculous to deserve even mention. These stones, artfully prepared,

> had been intentionally deposited in a mountain, which he was in the habit of exploring, purposely to dupe the enthusiastic collector. Unfortunately, the silly and cruel trick, succeeded in so far, as to occasion to him, who was the subject of it, so great a degree of mortification, as, it is said, shortened his days.

All components of the standard story line, complete with moral messages, have already fallen into place—the absurdity of the fossils, the gullibility of the professor, the personal tragedy of his undoing, and the two attendant lessons for aspiring young scientists: do not engage in speculation beyond available evidence, and do not stray from the empirical method of direct observation.

In this century's earlier and standard work on the history of geology (*The Birth and Development of the Geological Sciences,* published in 1934), Frank Dawson Adams provides some embellishments that had accumulated over the years, including the unforgettable story, for which not a shred of evidence has ever existed, that Beringer capitulated when he found his own name in Hebrew letters on one of his stones. Adams's verbatim "borrowing" of Parkinson's last line also illustrates another reason for invariance of the canonical tale: later retellings copy their material from earlier sources:

> Some sons of Belial among his students prepared a number of artificial fossils by moulding forms of various living or imaginary things in clay which was then baked hard and scattered in fragments about on the hillsides where Beringer was wont to search for fossils. . . . The distressing climax was reached, however, when later he one day found a fragment bearing his own name upon it. So great was his chagrin and mortification in discovering that he had been made the subject of a cruel and silly hoax, that he endeavored to buy up the whole edition of his work. In doing so he impoverished himself and it is said shortened his days.

Modern textbooks tend to present a caricatured "triumphalist" account in their "obligatory" introductory pages on the history of their discipline—the view that science marches inexorably forward from dark superstition toward the refining light of truth. Thus, Beringer's story tends to acquire the additional moral that his undoing at least had the good effect of destroying old nonsense about the inorganic or mysterious origin of fossils—as in this text for first-year students, published in 1961:

> The idea that fossils were merely sports of nature was finally killed by ridicule in the early part of the eighteenth century. Johann Beringer, a professor at the University of Würzburg, enthusiastically argued against the organic nature of fossils. In 1726, he published a paleontological work . . . which included drawings of many true fossils but also of objects that represented the sun, the moon, stars, and Hebraic letters. It was not till later, when Beringer found a "fossil" with his own name on it, that he realized that his students, tired of his teachings, had planted these "fossils" and carefully led him to discover them for himself.

A recent trip to Morocco turned my thoughts to Beringer. For several years, I have watched, with increasing fascination and puzzlement, the virtual "takeover" of rock shops throughout the world by striking fossils from Morocco—primarily straight-shelled nautiloids (much older relatives of the coiled and modern chambered nautilus) preserved in black marbles and limestones, and usually sold as large, beautifully polished slabs intended for table or dresser tops. I wondered where these rocks occurred in such fantastic abundance; had the High Atlas Mountains been quarried away to sea level? I wanted to make sure that Morocco itself still existed as a discrete entity and not only as disaggregated fragments, fashioning the world's coffee tables.

I discovered that most of these fossils come from quarries in the rocky deserts, well and due east of Marrakech, and not from the intervening mountains. I also learned something else that alleviated my fears about imminent dispersal of an entire patrimony. Moroccan rock salesmen dot the landscape in limitless variety—from young boys hawking a specimen or two at every hairpin turn on the mountain roads, to impromptu stands at every lookout point, to large and formal shops in the cities and towns. The aggregate volume of rock must be immense, but the majority of items offered for sale are either entirely phony or at least strongly "enhanced." My focus of interest shifted dramatically: from worrying about sources and limits to studying the ranges and differential expertises of a major industry dedicated to the manufacture of fake fossils.

I must judge some "enhancements" as quite clever—as when the strong ribs on the shell of a genuine ammonite are extended by carving into the smallest and innermost whorls and then "improved" in regular expression on the outer coil. But other "ammonites" have simply been carved from scratch on a smoothed rock surface, or even cast in clay and then glued into a prepared hole in the rock. Other fakes can only be deemed absurd—as in my favorite example of a wormlike "thing" with circles on its back, grooves on both sides, eyes on a head shield, and a double projection, like a snake's forked tongue, extending out in front. (In this case the forger, too clever by half, at least recognized the correct principle of parts and counterparts—for the "complete" specimen includes two pieces that fit together, the projecting "fossil" on one slab, and the negative impression on the other, where the animal supposedly cast its form into the surrounding sediment. The forger even carved negative circles and grooves into the counterpart image, although these impressions do not match the projecting, and supposedly corresponding, embellishments on the "fossil" itself.)

But one style of fakery emerges as a kind of "industry standard," as defined by constant repetition and presence in all shops. (Whatever the unique and personal items offered for sale in any shop, this *vin ordinaire* of the genre always appears in abundance.) These "standards" feature small (up to four or six inches in length) flattened stones with a prominent creature spread out in three dimensions on the surface. The fossils span a full range from plausible "trilobites," to arthropods (crabs, lobsters, and scorpions, for example) with external hard parts that might conceivably fossilize (though never in such complete exactitude), to small vertebrates (mostly frogs and lizards) with a soft exterior, including such delicate features as fingers and eyes that cannot be preserved in the geological record.

After much scrutiny, I finally worked out the usual mode of manufacture. The fossil fakes are plaster casts, often remarkably well done. (The lizard that I bought, as seen in the accompanying photograph, must have been cast from life, for a magnifying glass reveals the individual pores and scales of the skin.) The forger cuts a flat surface on a real rock and then cements the plaster cast to this substrate. (If you look carefully from the side, you can always make out the junction of rock and plaster.) Some fakes have been crudely confected, but the best examples match the color and form of rock to overlying plaster so cleverly that distinctions become nearly invisible.

A fake fossil reptile from a Moroccan rock shop. Done in plaster from a live cast and then glued to the rock.

When I first set eyes on these fakes, I experienced the weirdest sense of déjà vu, an odd juxtaposition of old and new that sent shivers of fascination and discomfort up and down my spine—a feeling greatly enhanced by a day just spent in the medina of Fez, the ancient walled town that has scarcely been altered by a millennium of surrounding change, where only mules and donkeys carry the goods of commerce, and where high walls, labyrinthine streets, tiny open shops, and calls to prayer, enhanced during the fast of Ramadan, mark a world seemingly untouched by time, and conjuring up every stereotype held by an uninformed West about a "mysterious East." I looked at these standard fakes, and I saw Beringer's *Lügensteine* of 1726. The two styles are so uncannily similar that I wondered, at first, if the modern forgers had explicitly copied the plates of the *Lithographiae Wirceburgensis*—a silly idea that I dropped as soon as I returned and consulted my copy of Beringer's original. But the similarities remain overwhelming. I purchased two examples—a scorpion of sorts and a lizard—as virtual dead ringers for Beringer's *Lügensteine,* and I present a visual comparison of the two sets of fakes, separated by 250 years and a different process of manufacture (carved in Germany, cast in Morocco). I only wonder if the proprietor believed my assurances, rendered in my best commercial French, that I was a

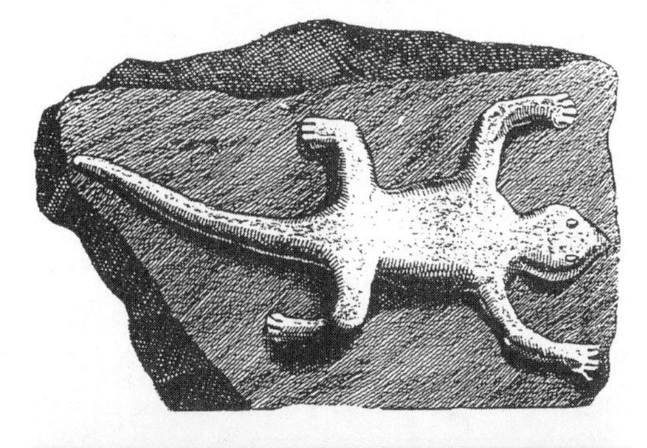

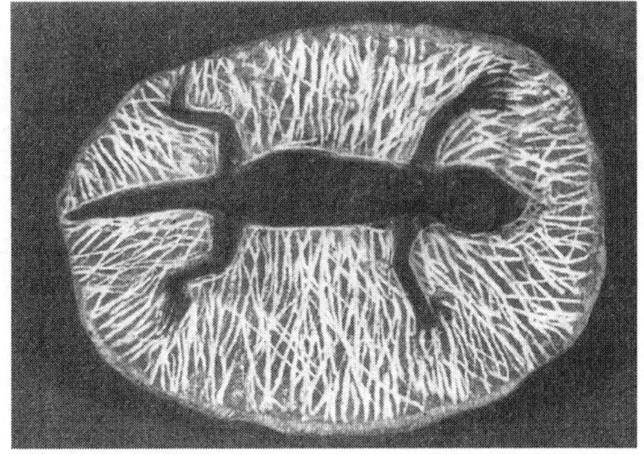

The striking similarity between the most famous fake in the history of paleontology (Beringer's Lügensteine, *or "lying stones," of 1726) and a modern Moroccan fabrication.*

professional paleontologist, and that his wares were *faux, absolument et sans doute*—or if he thought that I had just devised a bargaining tactic more clever than most.

But an odd similarity across disparate cultures and centuries doesn't provide a rich enough theme for an essay. I extracted sufficient generality only when I realized that this maximal likeness in appearance correlates with a difference in meaning that couldn't be more profound. A primary strategy of the experimental method in science works by a principle known since Roman times as *ceteris paribus* ("all other things being equal")—that is, if you wish to understand a controlling difference between two systems, keep all other features constant, for the difference may then be attributed to the only factor that you have allowed to vary. If, for example, you wish to test the effect of a new diet pill, try to establish two matched groups—folks of the same age, sex, weight, nutrition, health, habits, ethnicity, and so on. Then give the pill to one group and a placebo to the other (without telling the subjects what they have received, for such knowledge would, in itself, establish inequality based on differing psychological expectations). The technique, needless to say, does not work perfectly (for true *ceteris paribus* can never be obtained), but if the pill group loses a lot of weight, and the placebo group remains as obese as before, you may conclude that the pill probably works as hoped.

Ceteris paribus represents a far more distant pipe dream in trying to understand two different contexts in the developing history of a profession—for we cannot now manipulate a situation of our own design, but must study past circumstances in complex cultures not subject to regulation by our experimental ideals at all. But any constancy between the two contexts increases our hope of illustrating and understanding their variations in the following special way: if we examine the different treatment of the same object in two cultures, worlds apart, then at least we can attribute the observed variation to cultural distinctions, for the objects treated do not vary.

The effectively identical *Lügensteine* of early-eighteenth-century Würzburg and modern Marrakech embody such an interesting difference in proposed meaning and effective treatment by two cultures—and I am not sure that we should be happy about the contrast of then and now. But we must first correct the legend of Beringer and the original *Lügensteine* if we wish to grasp the essential difference.

As so often happens when canonical legends arise to impart moral lessons to later generations, the standard tale distorts nearly every important detail of Beringer's sad story. (I obtained my information primarily from an excellent

Note the exuberance and (by modern standards) whimsical nature of Beringer's fake fossils from 1726.

book published in 1963 by Melvin E. Jahn and Daniel J. Woolf, *The Lying Stones of Dr. Beringer,* University of California Press. Jahn and Woolf provide a complete translation of Beringer's volume, along with extensive commentary about the paleontology of Beringer's time. I used original sources from my own library for all quotations not from Beringer in this essay.)

First of all, on personal issues not directly relevant to the theme of this essay, Beringer wasn't tricked by a harmless student prank but rather purposely defrauded by two colleagues who hated his dismissive pomposity and wished to bring him down. These colleagues—J. Ignatz Roderick, professor of geography and algebra at the University of Würzburg, and Georg von Eckhart, librarian to the court and the university—"commissioned" the fake fossils (or, in Roderick's case, probably did much of the carving himself), and then hired a seventeen-year-old boy, Christian Zänger (who may also have helped with the carving), to plant them on the mountain. Zänger, a double agent of sorts, was then hired by Beringer (along with two other boys, both apparently innocent of the fraud) to excavate and collect the stones.

This information for revising the canonical tale lay hidden for two hundred years in the incomplete and somewhat contradictory records of hearings held in April 1726 before the Würzburg cathedral chapter and the city hall of Eivelstadt (the site of Beringer's mountain just outside Würzburg). The German scholar Heinrich Kirchner discovered these documents in 1934 in the town archives of Würzburg. These hearings focus on testimony of the three boys. Zänger, the "double agent," states that Roderick had devised the scheme because he "wished to accuse Dr. Beringer . . . because Beringer was so arrogant and despised them all." I was also impressed by the testimony of the two brothers hired by Beringer. Their innocence seems clear in the wonderfully ingenuous statement of Nicklaus Hahn that if he and his brother "could make such stones, they wouldn't be mere diggers."

The canonical tale may require Beringer's ruin to convey a desired moral, but the facts argue differently. I do not doubt that the doctor was painfully embarrassed, even mortified, by his exposed gullibility; but he evidently recovered, kept his job and titles, lived for another fourteen years, and published several more books (including, though probably not by his design or will, a posthumous second edition of his *Würzburg Lithography*!). Eckhart and Roderick, on the other hand, fell into well-earned disgrace. Eckhart died soon thereafter, and Roderick, having left Würzburg (voluntarily or not, we do not know), then wrote a humbling letter to the prince-bishop begging permission to return—which his grace allowed after due rebuke for Roderick's past deeds—and to regain access to the library and archives so that he could write a proper obituary for his deceased friend Eckhart.

But on the far more important intellectual theme of Beringer's significance in the history of paleontology, a different kind of correction inverts the conventional story in a particularly meaningful way. The usual cardboard tale of progressive science triumphant over past ignorance requires that benighted "bad guys," who upheld the old ways of theological superstition against objective evidence of observational science, be branded as both foolish and stubbornly unwilling to face nature's factuality. Since Beringer falls into this category of old and bad, we want to see him as hopelessly duped by preposterous fakes that any good observer should have recognized—hence the emphasis, in the canonical story, on Beringer's mortification and on the ridiculous character of the *Lügensteine* themselves.

The Würzburg carvings are, of course, absurd by modern definitions and understanding of fossils. We know that spiders' webs and lizards' eyes—not to mention solar rays and the Hebrew name of God—cannot fossilize, so the *Lügensteine* can only be human carvings. We laugh at Beringer for not making an identification that seems so obvious to us. But in so doing, we commit the greatest of all historical errors: arrogantly judging our forebears in the light of modern knowledge perforce unavailable to them. Of course the *Lügensteine* are preposterous, once we recognize fossils as preserved remains of ancient organisms. By this criterion, letters and solar emanations cannot be real fossils, and anyone who unites such objects with plausible images of organisms can only be a fool.

But when we enter Beringer's early-eighteenth-century world of geological understanding, his interpretations no longer seem so absurd. First of all, Beringer was puzzled by the unique character of his *Lügensteine,* and he adopted no dogmatic position about their meaning. He did regard them as natural and not carved (a portentous error, of course), but he demurred on further judgment and repeatedly stated that he had chosen to publish in order to provide information so that others might better debate the nature of fossils—a tactic that scientists supposedly value. We may regard the closing words of his penultimate chapter as a tad grandiose and self-serving, but shall we not praise the sentiment of openness?

> I have willingly submitted my plates to the scrutiny of wise men, desiring to learn their verdict, rather than to proclaim my own in this totally new and much mooted question. I address myself to scholars, hoping to be instructed by their most learned responses. . . . It is my fervent expectation that illustrious lithographers will shed light upon this dispute which is as obscure as it is unusual. I shall add thereto my own humble torch, nor shall I spare

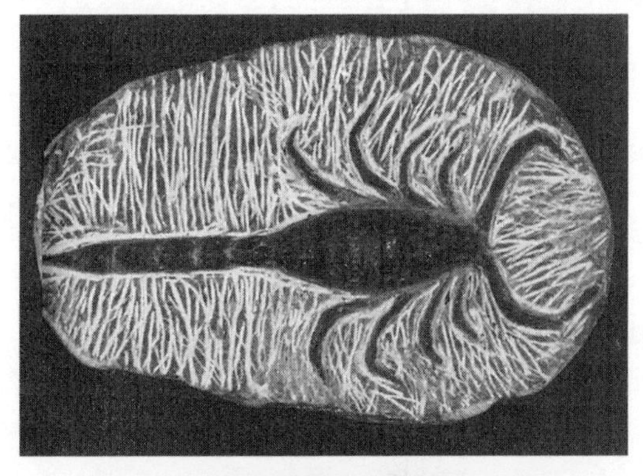

Another comparison between German fake fossils of 1726 and modern Moroccan fabrications.

> any effort to reveal and declare whatever future yields may rise from the Würzburg field under the continuous labors of my workers, and whatever opinion my mind may embrace.

More importantly, Beringer's hoaxers had not crafted preposterous objects but had cleverly contrived—for their purposes, remember, were venomous, not humorous—a fraud that might fool a man of decent will and reasonable intelligence by standards of interpretation then current. Beringer wrote his treatise at the tail end of a debate that had engulfed seventeenth-century science and had not yet been fully resolved: what did fossils represent, and what did they teach us about the age of the earth, the nature of our planet's history, and the meaning and definition of life?

Beringer regarded the *Lügensteine* as "natural" but not necessarily as organic in origin. In the great debate that he knew and documented so well, many scientists viewed fossils as inorganic products of the mineral realm that somehow mimicked the forms of organisms but might also take the shapes of other objects, including planets and letters. Therefore, in Beringer's world, the *Lügensteine* could not be dismissed as preposterous prima facie. This debate could not have engaged broader or more crucial issues for the developing sciences of geology and biology—for if fossils represent the remains of organisms, then the earth must be ancient, life must enjoy a long history of consistent change, and rocks must form from the deposition and hardening of sediments. But if fossils can originate as inorganic results of a "plastic power" in the mineral kingdom (that can fashion other interesting shapes like crystals, stalactites, and banded agates in different circumstances), then the earth may be young and virtually unchanged (except for the ravages of Noah's flood), while rocks, with their enclosed fossils, may be products of the original creation, not historical results of altered sediments.

If pictures of planets and Hebrew letters could be "fossils" made in the same way as apparent organisms, then the inorganic theory gains strong support—for a fossilized aleph or moonbeam could not be construed as a natural object deposited in a streambed and then fossilized when the surrounding sediment became buried and petrified. The inorganic theory had been fading rapidly in Beringer's time, while the organic alternative gained continually in support. But the inorganic view remained plausible, and the *Lügensteine* therefore become clever and diabolical, not preposterous and comical.

In Beringer's day, many scientists believed that simple organisms arose continually by spontaneous generation. If a polyp can originate by the influence of sunshine upon waters, or a maggot by heat upon decaying flesh, why not conjecture that simple images of objects might form upon rocks by natural interactions of light or heat upon the inherent "lapidifying forces" of the mineral kingdom? Consider, moreover, how puzzling the image of a fish *inside* a rock must have appeared to people who viewed these rocks as products of an original creation, not as historical outcomes of sedimentation. How could an organism get inside; and how could fossils be organisms if they frequently occur petrified, or made of the same stone as their surroundings? We now have simple and "obvious" answers to these questions, but Beringer and his colleagues still struggled—and any sympathetic understanding of early-eighteenth-century contexts should help us to grasp the centrality and excitement of these debates and to understand the *Lügensteine* as legitimately puzzling.

I do not, however, wish to absolve Beringer of all blame under an indefensibly pluralistic doctrine that all plausible explanations of past times may claim the same weight of judicious argument. The *Lügensteine* may not have been absurd, but Beringer had also encountered enough clues to uncover the hoax and avoid embarrassment. However, for several reasons involving flaws in character and passable intelligence short of true brilliance, Beringer forged on, finally trumping his judgment by his desire to be recognized and honored for a great discovery that had consumed so much of his time and expense. How could he relinquish the fame he could almost taste in writing:

> Behold these tablets, which I was inspired to edit, not only by my tireless zeal for public service, and by your wishes and those of my many friends, and by my strong filial love for Franconia, to which, from these figured fruits of this previously obscure mountain, no less glory will accrue than from the delicious wines of its vine-covered hills.

I am no fan of Dr. Beringer. He strikes me, first of all, as an insufferable pedant—so I can understand his colleagues' frustration, while not condoning their solutions. (I pride myself on always quoting from original sources, and I do own a copy of Beringer's treatise. I am no Latin scholar, but I can read and translate most works in this universal scientific language of Beringer's time. But I cannot make head or tail of the convoluted phrasings, the invented words, the absurdly twisted sentences of Beringer's prose, and I have had to rely on Jahn and Woolf's translation previously cited.)

Moreover, Beringer saw and reported more than enough evidence to uncover the hoax, had he been inclined to greater judiciousness. He noted that his *Lügensteine* bore no relationship to any other objects known to the burgeoning science of paleontology, not even to the numerous "real" fossils also found on his mountain. But instead of alerting him to possible fraud, these differences only fueled Beringer's hopes for fame. He made many observations that should have clued him in (even by standards of his own time) to the artificial carving of his fossils: why were they nearly always complete, and not usually fragmentary like most other finds; why did each object seem to fit so snugly and firmly on the enclosing rock; why did only the top sides protrude, while the lower parts merged with the underlying rock; why had letters and sunbeams not been found before; why did nearly all fossils appear in the same orientation, splayed out and viewed from the top, never from the side or bottom? Beringer's own words almost shout out the obvious and correct conclusion that he couldn't abide or even discern: "The figures expressed on these stones, especially those of insects, are so exactly fitted to the dimensions of the stones, that one would swear that they are the work of a very meticulous sculptor."

Beringer's arrogance brought him down in a much more direct manner as well. When Eckhart and Roderick learned that Beringer planned to publish his work, they realized that they had gone too far and became frightened. They tried to warn Beringer, by hints at first but later quite directly as their anxiety increased. Roderick even delivered some stones to Beringer and later showed his rival how they had been carved—hoping that Beringer would then draw an obvious inference about the rest of his identical collection.

Beringer, however, was now committed and would not be derailed. He replied with the argument of all true believers, the unshakable faith that resists all reason and evidence: yes, you have proven that *these* psychics are frauds, but *my* psychics are the real McCoy, and I must defend them even more strongly now that you have heaped unfair calumnies upon the entire enterprise. Beringer never mentions Eckhart and Roderick by name (so their unveiling awaited the 1934 discovery in the Würzburg town archives), but he had been forewarned of their activities. Beringer wrote in chapter 12 of his book:

> Then, when I had all but completed my work, I caught the rumor circulating throughout the city . . . that every one of these stones . . . was recently sculpted by hand, made to look as though at different periods they had been resurrected from a very old burial, and sold to me as to one indifferent to fraud and caught up in the blind greed of curiosity.

Beringer then tells the tale of Roderick's warning but excoriates his rival as an oafish modern caricature of Praxiteles (the preeminent Greek sculptor), out to discredit a great discovery by artificial mimicry:

> Our Praxiteles has issued, in an arrogant letter, a declaration of war. He has threatened to write a small treatise exposing my stones as supposititious [*sic*]—I should say, his stones, fashioned and fraudulently made by his hand. Thus does this man, virtually unknown among men of letters, still but a novice in the sciences, make a bid for the dawn of his fame in a shameful calumny and imposture.

If only Beringer had realized how truly and comprehensively he had spoken about "a shameful calumny and imposture." But Roderick succeeded because he had made his carvings sufficiently plausible to inspire belief by early-eighteenth-century standards. The undoing of all protagonists then followed because Beringer, in his overweening and stubborn arrogance, simply could not quench his ambition once a clever and plausible hoax had unleashed his ardor and vanity.

In summary, the *Lügensteine* of Würzburg played a notable role in the most important debate ever pursued in paleontology—a struggle that lasted for centuries and that placed the nature of reality itself up for grabs. By Beringer's time, this debate had largely been settled in favor of the organic nature of fossils, and this resolution would have occurred even if Beringer had never been born and the *Lügensteine* never carved. Beringer may have been a vain and arrogant man of limited talent, working in an academic backwater of his day, but at least he struggled with grand issues—and he fell because his hoaxers understood the great stakes and fashioned frauds that could be viewed as cleverly relevant to this intellectual battle, however preposterous they appear to us today with our additional knowledge and radically altered theories about the nature of reality and causation.

(One often needs a proper theory to set a context for the exposure of fraud. Piltdown Man fooled some of the world's best scientists for generations. I will never forget what W. E. le Gros Clark, one of the three scientists who exposed the fraud in the early 1950s, said to me when I asked him why the hoax had stood for forty years. Even an amateur in vertebrate anatomy—as this snail man can attest from personal experience—now has no trouble seeing the Piltdown bones for what they are. The staining is so crude, and the recent file marks on the orangutan teeth in the lower jaw so obvious—yet so necessary to make them look human in the forgers' plan, for the cusps of ape and human teeth differ so greatly. Le Gros Clark said to me: "One needed to approach the bones with the hypothesis of fraud already in mind. In such a context, the fakery immediately became obvious.")

The *Lügensteine* of Marrakech are, by contrast—and I don't know how else to say this—merely ludicrous and preposterous. No excuse save ignorance—and I do, of course, recognize the continued prevalence of this all-too-human trait—could possibly inspire a belief that the plaster blobs atop the Moroccan stones might be true fossils, the remains of ancient organisms. Beringer was grandly tricked in the pursuit of great truths, however inadequate his own skills. We are merely hoodwinked for a few dollars that mean little to most tourists but may make or break the lives of local carvers. *Caveat emptor.*

In contrasting the conflicting meanings of these identical fakes in such radically different historical contexts, I can only recall Karl Marx's famous opening line to *The Eighteenth Brumaire of Louis Napoleon,* his incisive essay on the rise to power of the vain and cynical Napoleon III after the revolution of 1848, in contrast with the elevated hopes and disappointments inspired by the original Napoleon. (The French revolutionary calendar had renamed the months and started time again at the establishment of the Republic. In this system, Napoleon's coup d'état occurred on the eighteenth of Brumaire, a foggy month in a renamed autumn, of year VIII—or November 9, 1799. Marx, now justly out of fashion for horrors later committed in his name, remains a brilliant analyst of historical patterns.) Marx opened his polemical treatise by noting that all great events in history occur twice—the first time as tragedy, and the second as farce.

Beringer was a pompous ass, and his florid and convoluted phrases represent a caricature of true scholarship. Still, he fell in the course of a great debate, using his limited talents to defend an inquiry that he loved and that even more pompous fools of his time despised—those who argued that refined people wouldn't dirty their hands in the muck of mountains but would solve the world's pressing issues under their wigs in their drawing rooms. Beringer characterized this opposition from the pseudo-elegant glitterati of his day:

> They pursue [paleontology] with an especially censorious rod, and condemn it to rejection from the world of erudition as one of the wanton futilities of intellectual idlers. To what purpose, they ask, do we stare fixedly with eye and mind at small stones and figured rocks, at little images of animals or plants, the rubbish of mountain and stream, found by chance amid the muck and sand of land and sea?

He then defended his profession with the greatest of geological metaphors:

> any [paleontologist], like David of old, would be able with one flawless stone picked from the bosom of Nature, to prostrate, by one blow

on the forehead, the gigantic mass of objections and satires and to vindicate the honor of this sublime science from all its calumniators.

Beringer, to his misfortune and largely as a result of his own limitations, did not pick a "flawless stone," but he properly defended the importance of paleontology and of empirical science in general. As a final irony, Beringer could not have been more wrong about the *Lügensteine,* but he couldn't have been more right about the power of paleontology. Science has so revolutionized our view of reality since 1726 that we, in our current style of arrogance, can only regard the Würzburg *Lügensteine* as preposterous, because we unfairly impose our modern context and fail to understand Beringer's world, including the deep issues that made his hoaxing a tragedy rather than a farce.

Our current reality features an unslayable Goliath of commercialism, and modern scientific Davids must make an honorable peace, for a slingshot cannot win this battle. I may be terribly old-fashioned (shades, I hope not, of poor Beringer)—but I continue to believe that such honor can only be sought in separation and mutual respect. Opportunities for increasing fusion with the world of commerce surround us with almost overwhelming temptation, for the immediate and palpable "rewards" are so great. So scientists go to work for competing pharmaceutical or computer companies, make monumental salaries, but cannot choose their topics of research or publish their work. And museums expand their gift shops to the size of their neglected exhibit halls, and purvey their dinosaurs largely for dollars in the form of images on coffee mugs and T-shirts, or by special exhibits, at fancy prices, of robotic models, built by commercial companies, hired for the show, and featuring, as their come-on, the very properties—mostly hideous growls and lurid colors—that leave no evidence in the fossil record and therefore remain a matter of pure conjecture to science.

I am relieved that Sue the *Tyrannosaurus,* sold at auction by Sotheby's for more than 8 million dollars, will go to Chicago's Field Museum and not to the anonymity of some corporate boardroom, to stand (perhaps) next to a phony Van Gogh. But I am not happy that no natural history museum in the world can pony up the funds for such a purpose—and that McDonald's had to provide the cash. McDonald's is not, after all, an eleemosynary institution, and they will legitimately want their piece for their price. Will it be the Happy Meal Hall of Paleontology at the Field Museum? (Will we ever again be able to view a public object with civic dignity, unencumbered by commercial messages? Must city buses be fully painted as movable ads, lampposts smothered, taxis festooned, even seats in concert halls sold one by one to donors and embellished in perpetuity with their names on silver plaques?) Or will we soon see Sue the

Robotic Tyrannosaur—the purchase of the name rather than the thing, for Sue's actual skeleton cannot improve the colors or sounds of the robots, and her value, in this context, lies only in the recognition of her name (and the memory of the dollars she attracted), not in her truly immense scientific worth.

I am neither an idealist nor a Luddite in this matter. I just feel that the world of commerce and the world of intellect, by their intrinsic natures, must pursue different values and priorities—while the commercial world looms so much larger than our domain that we can only be engulfed and destroyed if we make a devil's bargain of fusion for short-term gain. The worth of fossils simply cannot be measured in dollars. But the *Lügensteine* of Marrakech can only be assessed in this purely symbolic way—for the Moroccan fakes have no intellectual value and can bring only what the traffic (and human gullibility) will bear. We cannot possibly improve upon Shakespeare's famous words for this sorry situation—and this ray of hope for the honor and differences of intellect over cash:

> *Who steals my purse steals trash . . .*
> *But he that filches from me my good name*
> *Robs me of that which not enriches him,*
> *And makes me poor indeed.*

But we must also remember that these words are spoken by the villainous Iago, who will soon make Othello a victim, by exploiting the Moor's own intemperance, of the most poignant and tragic deception in all our literature. Any modern intellectual, to avoid Beringer's sad fate, must hold on to the dream—while keeping a cold eye on immediate realities. Follow your bliss, but remember that handkerchiefs can be planted for evil ends and fossils carved for ready cash.

2

The Sharp-Eyed Lynx, Outfoxed by Nature

I. Galileo Galilei and the Three Globes of Saturn

In 1603, Federico Cesi, the duke of Acquasparta, founded an organization that grew from uncertain beginnings to become the first scientific society in modern European history. Cesi (1585–1630), a teenaged nobleman, invited three slightly older friends (all in their mid-twenties) to establish the Accademia dei Lincei (Academy of the Lynxes), dedicated to scientific investigation ("reading this great, true, and universal book of the world," to cite Cesi's own words), and named for a sleek and wily carnivore, then still living in the forests of Italy and renowned in song and story for unparalleled sight among mammals.

The legend of the sharp-eyed lynx had arisen in ancient times and persisted to Cesi's day. Pliny's canonical compendium of

The official emblem of Europe's first scientific society, the Accademia dei Lincei (Academy of the Lynxes), founded in 1603 and including Galileo as an early member.

natural history had called the lynx "the most clear sighted of all quadrupeds." Plutarch had embellished the legend by speaking of "the lynx, who can penetrate through trees and rocks with its sharp sight." And Galen, ever the comparative anatomist, had written: "We would seem absurdly weak in our powers of vision if we compared our sight to the acuity of the lynx or the eagle." (I have translated these aphorisms directly from Conrad Gesner's 1551 compendium on mammals, the standard source for such information in Cesi's day.)

Still, despite Cesi's ambitious names and aims, the academy of four young men faltered at first. Cesi's father made a vigorous attempt to stop his son's foolishness, and the four Lynxes all dispersed to their native cities, keeping their organization alive only by the uncertain media of post and messages. But Cesi persevered and triumphed (for a time), thanks to several skills and circumstances. He acquired more power and prestige, both by growing up and by inheriting substantial wealth. Most importantly, he became a consummate diplomat and facilitator within the maximally suspicious and labyrinthine world of civil and ecclesiastical politics in Rome during the Counter-Reformation. The Lynxes flourished largely because Cesi managed to keep the suspicions of popes and cardinals at bay, while science prepared to fracture old views of the cosmos, and to develop radically new theories about the nature of matter and causation.

As a brilliant administrator, Cesi knew that he needed more clout among the membership of the Lynxes. He therefore recruited, as the fifth and sixth members of an organization that would eventually reach a roster of about thirty, two of the most prestigious thinkers and doers of early-seventeenth-century life. In 1610, he journeyed to Naples, where he persuaded the senior spokesman of the fading Neoplatonic school—the seventy-five-year-old Giambattista

Della Porta—to join a group of men young enough to be his grandsons. Then, in 1611, Cesi made his preeminent catch, when he recruited the hottest intellectual property in the Western world, Galileo Galilei (1564–1642), to become the sixth member of the Lynxes.

The year before, in 1610, Galileo had provided an ultimate proof for the cliché that good things come in small packages by publishing *Sidereus nuncius* (Starry messenger)—little more than a pamphlet really, but containing more oomph per paragraph than anything else ever achieved in the history of science or printing. Galileo shook the earth by turning his newly invented telescope upon the cosmos and seeing the moon as a planet with mountains and valleys, not as the perfect sphere required by conventional concepts of science and theology. Galileo also reported that thousands of previously invisible stars build the Milky Way, thus extending the cosmos beyond any previously conceivable limit; and that four moons orbit Jupiter, forming a miniature world analogous to the motion of planets around a central body. Moreover, Galileo pointed out, if satellites circle planets, then the crystalline spheres, supposedly marking the domain of each planet, and ordered as a set of concentric shells around the central earth, could not exist—for the revolution of moons would shatter these mystical structures of a geometrically perfect, unsullied, and unchanging cosmos, God's empyrean realm.

But Galileo also made some errors in his initial survey, and I have always been struck that standard books on the history of astronomy, written in the heroic or hagiographical mode, almost never mention (or relegate to an awkward footnote) the most prominent of Galileo's mistakes—for the story strikes me as fascinating and much more informative about the nature of science, and of creativity in general, than any of his valid observations.

Galileo also focused his telescope on Saturn, the most distant of the visible planets—and he saw the famous rings. But he could not properly visualize or interpret what he had observed, presumably because his conceptual world lacked the requisite "space" for such a peculiar object (while his telescope remained too crude to render the rings with enough clarity to force his mind, already benumbed by so many surprises, to the most peculiar and unanticipated conclusion of all).

The stymied Galileo looked and looked, and focused and focused, night after night. He finally interpreted Saturn as a threefold body, with a central sphere flanked by two smaller spheres of equal size, each touching the main planet. Following a common custom of the day—established to preserve claims of priority while not revealing preliminary conclusions ripe for theft by others—

Galileo encoded his interpretation as a Latin anagram, which he posted to his friend and leading compatriot in astronomical research, Johannes Kepler.

Kepler may have matched Galileo in brilliance, but he never resolved the anagram correctly, and he misinterpreted the message as a statement about the planet Mars. In frustration (and a bit of pique), he begged Galileo for the answer. His colleague replied with the intended solution:

Altissimum planetam tergeminum observavi.
[I have observed that the farthest planet is threefold.]

I regard the last word of Galileo's anagram as especially revealing. He does not advocate his solution by stating "I conjecture," "I hypothesize," "I infer," or "It seems to me that the best interpretation . . ." Instead, he boldly writes *"observavi"*—I have *observed.* No other word could capture, with such terseness and accuracy, the major change in concept and procedure (not to mention ethical valuation) that marked the transition to what we call "modern" science. An older style (as found, for example, in Gesner's compendium on mammals, cited above) would not have dishonored a claim for direct observation, but would have evaluated such an argument as a corroborative afterthought, surely secondary in weight to such criteria as the testimony of classical authors and logical consistency with a conception of the universe "known" to be both true and just—in other words, to authority and fixed "reasonableness."

But the new spirit of skepticism toward past certainty, coupled with respect for "pure" and personal observation—then being stressed by Francis Bacon in England, René Descartes in France, and the Lynxes in Italy—was sweeping through the intellectual world, upsetting all standard procedures of former times and giving birth to the modern form of an institution now called "science." Thus, Galileo supported his theory of Saturn with the strongest possible claim of the new order, the one argument that could sweep aside all opposition by claiming a direct, immediate, and unsullied message from nature. Galileo simply said: I have observed it; I have seen it with my own eyes. How could old Aristotle, or even the present pope himself, deny such evidence?

I do not intend, in this essay, to debunk the usual view that such a transition from old authority to direct observation marks a defining (and wonderfully salutary) event in the history of scientific methodology. But I do wish to note that all great mythologies include harmful simplicities amidst their genuine reforms—and that these negative features often induce the ironic consequence of saddling an original revolutionary doctrine with its own form of restrictive and unquestioned authority. The idea that observation can be pure and unsullied (and therefore beyond dispute)—and that great scientists are, by implication, people who can free their minds from the constraints of surrounding culture and reach conclusions strictly by untrammeled experiment and observation, joined with clear and universal logical reasoning—has often harmed science by turning the empiricist method into a shibboleth. The irony of this situation fills me with a mixture of pain for a derailed (if impossible) ideal and amusement for human foibles—as a method devised to undermine proof by authority becomes, in its turn, a species of dogma itself. Thus, if only to honor the truism that liberty requires eternal vigilance, we must also act as watchdogs to debunk the authoritarian form of the empiricist myth—and to reassert the quintessentially human theme that scientists can work only within their social and psychological contexts. Such an assertion does not debase the institution of science, but rather enriches our view of the greatest dialectic in human history: the transformation of society by scientific progress, which can only arise within a matrix set, constrained, and facilitated by society.

I know no better illustration of this central principle than the tale of Galileo's losing struggle with Saturn, for he insisted on validation by pure sight *(observavi),* and he could never see his quarry correctly—presumably because his intellectual domain included no option for rings around a planet. Galileo did not just "see" Saturn; he had to interpret an object in his lens by classifying an ambiguous shape (the best that his poor optics could provide) within the structure of his mental space—and rings didn't inhabit this interior world.

The great Dutch astronomer Christiaan Huygens finally recognized the rings of Saturn in 1656, more than a decade after Galileo's death. Galileo, who had wrestled mightily with Saturn, never moved beyond his trigeminal claim, and finally gave up and turned to other pursuits. In his 1613 book on sunspots, published by the Lynxes (with the author designated on the title page as Galileo Galilei Linceo), he continued to insist that Saturn must be threefold because he had so observed the planet: "I have resolved not to put anything around Saturn except what I have already observed and revealed—that is, two small stars which touch it, one to the east and one to the west." Against a colleague who interpreted the planet as oblong, Galileo simply asserted his superior vision. The colleague, Galileo wrote, had viewed Saturn less often and with a much poorer telescope, "where perfection is lacking, [and] the shape and distinction of the three stars imperfectly seen. I, who have observed it a thousand times at different periods with an excellent instrument, can assure you that no change whatever is to be seen in it."

Yet just as Galileo prepared his book on sunspots for publication, he observed Saturn again after a hiatus of two years—and the two side planets had disappeared (a situation produced, we now know, when the planet's changing orientation presents the rings directly on edge—that is, as an invisible line in Galileo's poor telescope). The stunned Galileo, reduced to a most uncharacteristic modesty, had just enough time to make an addition to the last chapter of his book. He abjured nothing about his previous observations or about the righteousness of the empirical method in general. He merely confessed his puzzlement, making a lovely classical allusion to the primary myth about the planet's eponym:

> I had discovered Saturn to be three-bodied. . . . When I first saw them they seemed almost to touch, and they remained so for almost two years without the least change. It was reasonable to believe them to be fixed. . . . Hence I stopped observing Saturn for more than two years. But in the past few days I returned to it and found it to be solitary, without its customary supporting stars, and as perfectly round and sharply bounded as Jupiter. Now what can be said of this strange metamorphosis? That the two lesser stars have been consumed? . . . Has Saturn devoured his children? Or was it indeed an illusion and a fraud with which the lenses of my telescope deceived me for so long—and not only me, but many others who have observed it with me? . . . I need not say anything definite upon so strange and unexpected an event; it is too recent, too unparalleled, and I am restrained by my own inadequacy and the fear of error.

After this lengthy preamble on the maximally celebrated Galileo, let me now present the main subject of this essay: the virtually unknown Francesco Stelluti, one of the original four Lynxes, a loyal friend and supporter of Galileo, and the man who tried to maintain, and eventually disbanded with dignity (in 1652), the original Academy of the Lynxes, fatally weakened after Cesi's untimely death in 1630. The previously uncharted links between Stelluti and Galileo are rich and fascinating (I would have said "the links between these Lynxes," if the pun were not so egregious), and these connections provide a poignant illustration of this essay's central theme: the power and poverty of pure empiricism, and the need to scrutinize social and intellectual contexts, both for practicing scientists (so they will not be beguiled) and for all people who wish to understand the role and history of knowledge (so they will grasp the necessary and complex interdigitation of science and society).

The original Lynxes began with all the bravado and secrecy of a typical boys' club (Cesi, remember, was only eighteen years old, while his three compatriots were all twenty-six). They wrote complex rules and enunciated lofty ideals. (I do not know whether or not they developed a secret handshake!) Each adopted a special role, received a Latin moniker, and took a planet for his emblem. The leader Cesi commanded the botanical sciences as Coelivagus (the heavenly wanderer); the Dutchman Johannes van Heeck would read and interpret classical philosophy as Illuminatus; Anastasio de Filiis became the group's historian and secretary as Eclipsatus. Poor Francesco Stelluti, who published little and evidently saw himself as a systematic plodder, took up mathematics and geometry under the name of Tardigradus (the slow stepper). For his planet, Stelluti received the most distant and most slowly revolving body—Saturn, the subject of Galileo's error!

In their maturity, the Lynxes would provide powerful intellectual and institutional support for the open and empirical approach to science, as promoted by their most prominent member, Galileo. But at their beginnings, as a small club of young men, the Lynxes preferred the older tradition of science as an arcane and secret form of knowledge, vouchsafed only to initiates who learned the codes and formulae that could reveal the mysterious harmonies of universal order—the astrological links between planetary positions and human lives; the alchemical potions and philosopher's stones, heated in vats that could transmute base metals to gold (double, double toil and trouble; fire burn and cauldron bubble, to cite some famous witches); and the experiments in smoke, mirrors, and optical illusions that occupied an uncertain position between categories now labeled as magic and science, but then conflated. Giambattista Della Porta, the fifth Lynx, had survived as a living legend of this fading philosophy. Della Porta had made his reputation in 1558, long before the birth of any original Lynx, with a book entitled *Magia naturalis,* or *Natural Magic.* As a young man in Naples, Della Porta had founded his own arcane organization, the Accademia dei Segreti (Academy of Secrets), dedicated to alchemical and astrological knowledge, and later officially suppressed by the Inquisition.

By initiating the aged Della Porta into the Academy of Lynxes, Cesi and his compatriots showed the strength of their earlier intellectual allegiances. By inducting Galileo the next year, they displayed their ambivalence, and their growing attraction to a new view of knowledge and scientific procedure.

The election of both newcomers virtually guaranteed a period of definitional struggle within the Academy, for no love could unite Della Porta and Galileo, who not only differed maximally in their basic philosophical approaches to science, but also nearly came to blows for a much more specific reason vested in the eternally contentious issue of priority. Galileo never claimed that he had invented the telescope from scratch. He stated that he had heard reports about a crude version during a trip to Venice in 1609. He recognized the optical principles behind the device, and then built a more powerful machine that could survey the heavens. But Della Porta, who had used lenses and mirrors for many demonstrations and illusions in his *Natural Magic,* and who surely understood the rules of optics, then claimed that he had formulated all the principles for building a telescope (although he had not constructed the device) and therefore deserved primary credit for the invention. Although tensions remained high, the festering issue never erupted into overt battle because Galileo and Della Porta held each other in mutual respect, and Della Porta died in 1615 before any growing bitterness could bubble over.

Stelluti first encountered Galileo in the context of this struggle—and he initially took Della Porta's side! In 1610, with Della Porta inscribed as a Lynx but Galileo not yet a member, Stelluti wrote a gossipy letter to his brother about the furor generated by *Sidereus nuncius* and the dubious claims of the pamphlet's author:

> I believe that by now you must have seen Galileo, he of the *Siderius nuncius* . . . Giambattista Della Porta wrote about [the telescope] more than thirty years ago in his *Natural Magic* . . . so poor Galileo will be besmirched. But, nonetheless, the Grand Duke has given him 800 piastres.*

But when Galileo joined the Lynxes, and as his fame and success solidified and spread, Stelluti and his compatriots muted their suspicions and eventually became fervent Galileans. With Della Porta dead and *Starry Messenger* riding a truly cosmic crest of triumph, the Academy of the Lynxes grew to become Galileo's strongest intellectual (and practical) base, the primary institutional supporters of the new, open, empirical, and experimental view of scientific knowledge. Making the link between Galileo's error and Stelluti's emblem,

---

*The quotations from Galileo's *Letters on Sunspots* come from Stillman Drake's 1957 English translation, published by Anchor Books. I have translated all other quotes from the Italian of Stelluti's 1637 monograph on fossil wood, letters from several volumes of the *Edizione Nazionale* of Galileo's complete works, and three standard sources on the history of the Academy of the Lynxes: *Breve storia della Accademia dei Lincei* by D. Carutti (Rome: Salviucci, 1883); *Contributi alla storia della Accademia dei Lincei* by G. Gabrieli (Rome, 1989); and *L'Accademia dei Lincei e la cultura europea nel XVII secolo,* a catalog for a traveling exhibit about the Lynxes by A. M. Capecchi and several other authors, published in 1991.

Cesi wrote to Stelluti in 1611 about the wonders of the telescope, as revealed by Galileo himself, then paying a long visit to the duke of Acquasparta:

> Each evening we see new things in the heavens, the true office of the Lynxes. Jupiter and its four revolving satellites; the moon with its mountains, caverns, and rivers; the horns of Venus; and Saturn, your own triple-star [*il triplice suo Saturno*].

Such floods of reforming novelty tend to alienate reigning powers, to say the least—a generality greatly exacerbated in early-seventeenth-century Rome, where the papal government, besieged by wars and assaulted by the successes of the Reformation, felt especially unfriendly to unorthodoxy of any sort. Galileo had written a first note of cautious support for the Copernican system at the end of his *Letters on Sunspots* (published by the Lynxes in 1613). Soon afterward, in 1616, the Church officially declared the Copernican doctrine false and forbade Galileo to teach heliocentrism as a physical reality (though he could continue to discuss the Copernican system as a "mathematical hypothesis"). Galileo kept his nose clean for a while and moved on to other subjects. But then, in 1623, the Lynxes rejoiced in an unanticipated event that Galileo called a "great conjuncture" *(mirabel congiuntura):* the elevation of his friend and supporter Maffeo Barberini to the papacy as Urban VIII. (In an act of literal nepotism, Maffeo quickly named his nephew Francesco Barberini as his first new cardinal. In the same year of 1623, Francesco Barberini became the twenty-ninth member of the Lynxes.)

On August 12, 1623, Stelluti wrote from Rome to Galileo, then in Florence, expressing both his practical and intellectual joy in the outcome of local elections. Three members of the Lynxes would be serving in the new papal government, along with "many other friends." Stelluti then enthused about the new boss:

> The creation of the new pope has filled us all with rejoicing, for he is a man of such valor and goodness, as you yourself know so well. And he is a particular supporter of learned men, so we shall have a supreme patron. . . . We pray to the Lord God to preserve the life of this pope for a long time.

The Lynxes, suffused with hope that freedom of scientific inquiry would now be established, met for an extended convention and planning session at

Cesi's estate in 1624. Galileo had just built the first usable microscope for scientific investigation, after recognizing that lenses, properly arranged, could magnify truly tiny nearby objects, as well as enormous cosmic bodies rendered tiny in appearance by their great distance from human observers. Anticipating the forthcoming gathering of the Lynxes, Galileo sent one of his first microscopes to Cesi, along with a note describing his second great optical invention:

> I have examined a great many tiny animals with infinite admiration. Mosquitoes are the most horrible among them. . . . I have seen, with great contentment, how flies and other tiny animals can walk across mirrors, and even upside down. But you, my lord, will have a great opportunity to view thousands and thousands of details. . . . In short, you will be able to enjoy infinite contemplation of nature's grandness, and how subtly, and with what incredible diligence, she works.

Galileo's microscope entranced the Lynxes and became the hit of their meeting. Stelluti took a special interest and used the new device to observe and draw the anatomy of bees. In 1625, Stelluti published his results, including a large engraving of three bees drawn under Galileo's instrument. Historian of science Charles Singer cites these bees as "the earliest figures still extant drawn with the aid of the microscope." If the name of the sadly underrated Francesco Stelluti, the tardigrade among the Lynxes, has survived at all in conventional annals of the history of science, he perseveres only as an entry in the "list of firsts" for his microscopical drawing.

The Lynxes, always savvy as well as smart, did not choose to draw bees for abstract amusement. Not coincidentally, the family crest of Maffeo Barberini, the new pope and the Lynxes' anticipated patron, featured three bees. Stelluti dedicated his work to Urban VIII, writing in a banner placed above the three bees: "To Urban VIII Pontifex Optimus Maximus . . . from the Academy of the Lynxes, and in perpetual devotion, we offer you this symbol."

The emboldened Galileo now decided to come out of intellectual hiding, and to risk a discussion of the Copernican system. In 1632, he published his epochal masterpiece in the history of science and, from the resulting tragedy, the history of society as well: *Dialogo . . . sopra i due massimi sistemi nel mondo tolemaico e copernicano* (A dialogue on the two great systems of the world, Ptolemaic and Copernican). Galileo hoped that he could avoid any ecclesiastical trouble by framing the work as a dialogue—an argument between a sup-

The first published scientific figure based on observations under a microscope—Stelluti's 1625 image of bees, drawn to honor the new pope, Urban VIII, whose family crest featured three bees.

porter of the earth-centered Ptolemaic system and a partisan of Copernicus's sun-centered view.

We all know the tragic outcome of this decision only too well. The pope, Galileo's erstwhile friend, became furious and ordered the scientist to stand trial before the Roman Inquisition. This tribunal convicted Galileo and forced him to abjure, on his knees, his "false" and heretical Copernican beliefs. The Inquisition then placed him under a form of house arrest for the remainder of his life, on his small estate at Arcetri. Galileo's situation did not resemble solitary confinement at Alcatraz, and he remained fully active in scientific affairs by receiving visitors and engaging in voluminous correspondence to the moment of his death (even though blindness afflicted his last four years). In 1638, and partly by stealth, Galileo wrote his second great book in dialogue form (with the same protagonists) and had a copy smuggled to the Netherlands for publication: *Discourses and Mathematical Demonstrations Concerning Two New*

The famous frontispiece of Galileo's dialogue between Ptolemy and Copernicus (with Aristotle as their mediator).

Sciences. But he was not allowed to leave Arcetri either, as the vindictive pope, still feeling betrayed, refused Galileo's requests to attend Easter mass and to consult doctors in Florence when his sight began to fail.

The literature on the whys and wherefores of Galileo's ordeal could fill a large room in a scholarly library, and I shall not attempt even the barest summary here. (The most interesting and original of recent books include Mario Biagioli's *Galileo, Courtier,* University of Chicago Press, 1993; and Pietro Redondi's *Galileo Heretic,* Princeton University Press, 1987.) All agree that Galileo might have avoided his fate if any one of a hundred circumstances had unfolded in a slightly different manner. He was, in other words, a victim of bad luck and bad judgment (on both sides), not an inevitable sacrificial lamb in an eternal war between science and religion.

Until doing research for this essay, however, I had never appreciated the strength of one particularly relevant factor along the string of contingencies. From the vantage point of the Lynxes, Galileo would almost surely have managed to weave a subtle path around potential trouble, if the most final of all events had not intervened. In 1630, at age forty-five and the height of his influence, Federico Cesi, founder and perpetual leader of the Lynxes, died. Galileo learned the sad news in a letter from Stelluti: "My dear signor Galileo, with a trembling hand, and with eyes full of tears [*con man tremante, e con occhi pieni di lacrime*—such lamentation sounds so much better in Italian!], I must tell you the unhappy news of the loss of our leader, the duke of Acquasparta, as the result of an acute fever."

I feel confident that Cesi could have intervened to spare Galileo for two reasons. First, his caution and diplomacy, combined with his uncanny sense of the practical, would have suppressed Galileo's famous and fatal impetuosity. Galileo, ever testing the limits, ever pushing beyond into a realm of danger, did cast his work in the form of a dialogue between a Copernican and a supporter of Ptolemy's earth-centered universe. But he had scarcely devised a fair fight. The supporter of Ptolemy bore the name Simplicio, and the quality of his arguments matched his moniker. Moreover, Urban VIII developed a sneaking suspicion that Simplicio might represent a caricature of his own imperial self—hence his angry feeling that Galileo had betrayed an agreement to discuss Copernicanism as a coherent theory among equally viable alternatives. If Cesi had lived, he would, no doubt, have insisted that Galileo write his dialogue in a less partisan, or at least a more subtly veiled, form. And Cesi would have prevailed, both because Galileo respected his judgment so highly, and because the Lynxes intended to publish his book at Cesi's expense.

Second, Cesi operated as one of the most consummate politicians on the Roman scene. As a diplomat and nobleman (contrasted with Galileo's status as a commoner and something of a hothead), Cesi would have greased all the wheels and prepared a smooth way. Galileo recognized the dimensions of his personal misfortune only too well. He wrote to his friend G. B. Baliani in 1630, just before Cesi's death:

> I was in Rome last month to obtain a license to print the Dialogue that I am writing to examine the two great systems, Ptolemaic and Copernican. . . . Truly, I would have left all this in the hands of our most excellent prince Cesi, who would have accomplished it with much care, as he has done for my other works. But he is feeling indisposed, and now I hear that he is worse, and may be in danger.

Cesi's death produced two complex and intertwined results lying at the heart of this essay: the subsequent, and preventable, condemnation of Galileo; and the attrition and inevitable extinction of the Society of Lynxes. Stelluti tried valiantly to keep the Lynxes alive. He importuned Francesco Barberini, the cardinal nephew of the pope, and the only member of the Lynxes with enough clout to fill Cesi's shoes, to become the new leader. Barberini's refusal sealed the Lynxes' fate, for no other sufficiently rich and noble patron could be found. Cesi soldiered on for a while and, in a noble last hurrah, finally published, in 1651, the volume on the natural history of the New World that the Lynxes had been planning for decades: *Nova plantarum et mineralium mexicanorum historia.* In a final loving tribute, Stelluti included Cesi's unpublished work on botanical classification in an appendix. In 1652, Stelluti, the last original Lynx, died—and the organization that he had nourished for a lifetime, in his own slow and steady manner, ceased to exist.

II. Francesco Stelluti and the Mineral Wood of Acquasparta

Francesco Stelluti remained faithful to Galileo during his friend's final years of internal exile and arrest. On November 3, 1635, he wrote a long and interesting letter to Galileo at Arcetri, trying to cheer his friend with news from the world of science. Stelluti first expressed his sympathy for Galileo's plight: "God knows how grieved and pained I am by your ordeal" *(Dio sa quanto mi son doluto e doglio de' suoi travagli).* Stelluti then attempted to raise Galileo's spirits with the latest report on an old project of the Lynxes—an analysis of some curious fossil wood found on Cesi's estate:

> You should know that while I was in Rome, Signor Cioli visited the Duchess [Cesi's widow] several times, and that she gave him, at his departure, several pieces of the fossil wood that originates near Acquasparta. . . . He wanted to know where it was found, and how it was generated . . . for he noted that Prince Cesi, of blessed memory, had planned to write about it. The Duchess then asked me to write something about this, and I have done so, and sent it to Signor Cioli, together with a package of several pieces of the wood, some petrified, and some just beginning to be petrified.

This fossil wood had long vexed and fascinated the Lynxes. Stelluti had described the problem to Galileo in a letter of August 23, 1624, written just before the Lynxes' convention and the fateful series of events initiated by Stelluti's microscopical drawings of bees, intended to curry favor with the new pope.

> Our lord prince [Cesi] kisses your hands and is eager to hear good news from you. He is doing very well, despite the enervating heat, which does not cause him to lose any time in his studies and most beautiful observations on this mineralized wood. He has discovered several very large pieces, up to eleven palms [of the human hand, not the tree of the same name] in diameter, and others filled with lines of iron, or a material similar to iron. . . . If you can stop by here on your return to Florence, you can see all this wood, and where it originates, and some of the nearby mouths of fire [steaming volcanic pits near Acquasparta that played a major role in Stelluti's interpretation of the wood]. You will observe all this with both surprise and enthusiasm.

We don't usually think of Galileo as a geologist or paleontologist, but his catholic (with a small *c*!) interests encompassed everything that we would now call science, including all of natural history. Galileo took his new telescope to his first meeting of 1611 with Cesi and the Lynxes, and the members all became enthralled with Galileo's reconstructed cosmos. But he also brought, to the same meeting, a curious stone recently discovered by some alchemists in Bologna, called the *lapis Bononensis* (the Bologna stone), or the "solar sponge"—for the rock seemed to absorb, and then reflect, the sun's light. The specimens have been lost, and we still cannot be certain about the composition or the nature of Galileo's stone (found in the earth or artificially made). But we do know that the Lynxes became entranced by this geological wonder. Cesi,

A comparison of title pages for Galileo's book on sunspots and Stelluti's treatise on fossil wood, with both authors identified as members of the Lynx society.

TRATTATO
DEL LEGNO FOSSILE MINERALE
NVOVAMENTE SCOPERTO
NEL QVALE BREVEMENTE SI ACCENNA LA VARIA,
& mutabil natura di detto Legno, rappresentatoui con alcune
figure, che mostrano il luogo doue nasce, la diuersità
dell'onde, che in esso si vedono, e le sue così varie,
e marauigliose forme.
DI FRANCESCO STELLVTI ACCAD. LINCEO
DA FABRIANO.
All'Emin.mo & Reuer.mo Sig. Card.
FRANCESCO
BARBERINO.
IN ROMA, Appresso Vitale Mascardi, MDCXXXVII.
CON LICENZA DE'SVPERIORI.

ISTORIA
E DIMOSTRAZIONI
INTORNO ALLE MACCHIE SOLARI
E LORO ACCIDENTI
COMPRESE IN TRE LETTERE SCRITTE
ALL'ILLVSTRISSIMO SIGNOR
MARCO VELSERI LINCEO
DVVMVIRO D'AVGVSTA
CONSIGLIERO DI SVA MAESTA CESAREA
DAL SIGNOR
GALILEO GALILEI LINCEO
Nobil Fiorentino, Filosofo, e Matematico Primario del Serenis.
D. COSIMO II. GRAN DVCA DI TOSCANA.
Si aggiungono nel fine le Lettere, e Disquisizioni del finto Apelle.
IN ROMA, Appresso Giacomo Mascardi. MDCXIII.
CON LICENZA DE' SVPERIORI.

committed to a long stay at his estate in Acquasparta, begged Galileo for some specimens, which arrived in the spring of 1613. Cesi then wrote to Galileo: "I thank you in every way, for truly this is most precious, and soon I will enjoy the spectacle that, until now, absence from Rome has not permitted me" (I read this quotation and information about the Bologna stone in Paula Findlen's excellent book, *Processing Nature,* University of California Press, 1994).

Galileo then took a reciprocal interest in Cesi's own geological discovery—the fossil wood of Acquasparta; so Stelluti's letters reflect a clearly shared interest. Cesi did not live to publish his controversial theories on this fossil wood. Therefore, the ever-loyal Stelluti gathered the material together, wrote his own supporting text, engraved thirteen lovely plates, and published his most influential work (with the possible exception of those earlier bees) in 1637: *Trattato del legno fossile minerale nuovamente scoperto, nel quale brevemente si accenna la varia e mutabil natura di detto legno, rappresentatovi con alcune figure, che mostrano il luogo dove nasce, la diversita dell'onde, che in esso si vedono, e le sue cosi varie, e maravigliose forme*—a title almost as long as the following text (Treatise on newly discovered fossil mineralized wood, in which we point out the variable and mutable nature of this wood, as represented by several figures, which show the place where it originates, the diversity of waves [growth lines] that we see in it, and its highly varied and marvelous forms).

The title page illustrates several links with Galileo. Note the similar design and same publisher (Mascardi in Rome) for the two works. Both feature the official emblem of the Lynxes—the standard picture of the animal (copied from Gesner's 1551 compendium), surrounded by a laurel wreath and topped by the crown of Cesi's noble family. Both authors announce their affiliation by their name—the volume on sunspots by Galileo Galilei Linceo, the treatise on fossil wood by Francesco Stelluti Accad. Linceo. The ghosts of Galileo's tragedy also haunt Stelluti's title page, for the work bears a date of 1637 (lower right in Roman numerals), when Galileo lived in confinement at Arcetri, secretly writing his own last book. Moreover, Stelluti dedicates his treatise quite obsequiously "to the most eminent and most revered Signor Cardinal Francesco Barberini" (in type larger than the font used for Stelluti's own name), the nephew of the pope who had condemned Galileo, and the man who had refused Stelluti's invitation to lead (and save) the Lynxes after Cesi's death.

But the greatest and deepest similarity between Galileo's book on sunspots and Stelluti's treatise on fossil wood far transcends any visual likenesses, and resides instead in the nature of a conclusion, and a basic style of rhetoric and scientific procedure. Galileo presented his major discussion of Saturn in his book on sunspots (as quoted earlier in this essay)—where he stated baldly that an entirely false interpretation must be correct because he had observed the phenomenon with his own eyes. Stelluti's treatise on fossil wood presents a completely false (actually backward) interpretation of Cesi's discovery, and then uses exactly the same tactic of arguing for the necessary truth of his view because he had personally observed the phenomena he described!

Despite some practical inconveniences imposed by ruling powers committed neither to democracy nor to pluralism—one might, after all, end up burned like Bruno, or merely arrested, tried, convicted, and restricted like Galileo—the first half of the seventeenth century must rank as an apex of excitement for scientists. The most fundamental questions about the structure, meaning, and causes of natural phenomena all opened up anew, with no clear answers apparent, and the most radically different alternatives plausibly advocated by major intellects. By inventing a simple device for closer viewing, Galileo fractured the old conception of nature's grandest scale. Meanwhile, on earth, other scientists raised equally deep and disturbing questions about the very nature of matter and the basic modes of change and causality.

The nascent science of paleontology played a major role in this reconstruction of reality—primarily by providing crucial data to resolve the two debates that convulsed (and virtually defined) the profession in Stelluti and Galileo's time (see chapter 1 for more details on this subject):

1. What do fossils represent? Are they invariably the remains of organisms that lived in former times and became entombed in rocks, or can they be generated inorganically as products of formative forces within the mineral kingdom? (If such regular forms as crystals, and such complex shapes as stalactites, can arise inorganically, why should we deny that other petrified bodies, strongly resembling animals and plants, might also originate as products of the mineral kingdom?)

2. How shall we arrange and classify natural objects? Is nature built as a single continuum of complexity and vitality, a chain of being rising without a gap from dead and formless muds and clays to the pinnacle of humanity, perhaps even to God himself? Or can natural objects be placed into sharply separated, and immutably established, realms, each defined by such different principles of structure that no transitional forms between realms could even be imagined? Or in more concrete terms: does the old tripartite division of mineral, vegetable, and animal represent three loosely defined domains within a single continuum (with transitional forms between each pair), or a set of three utterly disparate modes, each serving as a distinct principle of organization for a unique category of natural objects?

Cesi had always argued, with force and eloquence, that the study of small objects on earth could yield as much reform and insight as Galileo's survey of the heavens. The microscope, in other words, would be as valuable as the telescope. Cesi wrote:

> If we do not know, collect, and master the smallest things, how will we ever succeed in grasping the large things, not to mention the biggest of all? We must invest our greatest zeal and diligence in the treatment and observation of the smallest objects. The largest of fires begins with a small spark; rivers are born from the tiniest drops, and grains of sand can build a great hill.

Therefore, when Cesi found a puzzling deposit of petrified wood near his estate, he used these small and humble fossils to address the two great questions outlined above—and he devised the wrong answer for each! Cesi argued that his fossil wood had arisen by transformation of earths and clays into forms resembling plants. His "wood" had therefore been generated from the mineral kingdom, proving that fossils could form inorganically. Cesi then claimed that his fossils stood midway between the mineral and vegetable kingdoms, providing a smooth bridge along a pure continuum. Nature must therefore be constructed as a chain of being. (Cesi had strongly advocated this position for a long time, so he can scarcely be regarded as a dispassionate or disinterested observer of fossils. His botanical classification, eventually published by Stelluti in 1651, arranged plants in a rising series from those he interpreted as most like minerals to forms that he viewed as most like animals.) Since Cesi could not classify his fossils into any conventional kingdom, he awarded them a separate name for a novel realm between minerals and plants—the Metallophytes.

Stelluti, playing his usual game of follow the leader, devoted his 1637 treatise to supporting Cesi's arguments for the transitional status of metallophytes and their origin from the mineral kingdom as transmuted earths and clays. The fossils may look like plants, but they originate from heated earths of the surrounding countryside (where subterranean magmas boil the local waters, thus abetting the conversion of loose earth to solid metallophyte). Stelluti concludes:

> The generation of this wood does not proceed from the seed or root of any plant, but only from a kind of earth, very much like clay, which little by little becomes transmuted to wood. Nature operates in such a manner until all this earth is converted into that kind of wood. And I believe that this occurs with the aid of heat from subterranean fires, which are found in this region.

To support this conclusion, Stelluti presented the following five basic arguments:

1. The fossil wood, generated from earth, only assumes the forms of tree trunks, never any other parts of true plants:

> It is clear that this wood is not born from seeds, roots or branches, like other plants, because we never find pieces of this wood with roots, or branches, or nerves [internal channels for fluids], as in other [truly vegetable] wood and trees, but only simple trunks of varied form.

Three figures of fossil wood from Stelluti's treatise of 1637.

2. The fossil trunks are not rounded, as in true trees, but rather compressed to oval shapes, because they grow *in situ* from earths flattened by the weight of overlying sediments (see the accompanying reproduction of Stelluti's figure):

> I believe that they adopt this oval shape because they must form under a great mass of earth, and cannot grow against the overlying weight to achieve the circular, or rather cylindrical, form assumed by the trunks of true trees. Thus, I can securely affirm that the original material of this wood must have been earth of a clayey composition.

3. Five of Stelluti's plates present detailed drawings of growth lines in the fossil wood (probably done, in part, with the aid of a microscope). Stelluti's argument for these inner details of structure follows his claim for the outward form of entire specimens: the growth lines form wandering patterns reflecting irregular pathways of generation from earth, following limits imposed by the weight of overlying sediments. These lines never form in regular concentric circles, as in true trees. Stelluti therefore calls them *onde,* or "waves," rather than growth lines:

> The waves and veins are not continuous, all following the same form, as in [vegetable] wood, but are shaped in a variety of ways—

> some long and straight, others constricted, others thick, others contorted, others meandering. . . . This mineral wood takes its shape from the press of the surrounding earth, and thus it has waves of such varied form.

4. In the argument that he regarded as most decisive, Stelluti held that many specimens can be found in the process of transition, with some parts still made of formless earth, others petrified in the shapes of wood, and still others fully converted to wood. Stelluti views these stages as an actual sequence of transformation. He writes about a large specimen, exposed *in situ:*

> In a ditch, we discovered a long layer of this wood . . . rather barrel-shaped, with one segment made of pure earth, another of mixed earth and wood, and another of pure wood. . . . We may therefore call it earth-wood *(creta legno).*

Later, he draws a smaller specimen (reproduced here from Stelluti's figure) and states:

> The interior part is made of wood and metal together, but the crust on the outside seems to be made of lateritic substance, that is, of terra cotta, as we find in bricks.

5. In a closing (and conclusive) flourish for the empirical method, Stelluti reports the results of a supposed experiment done several years before:

> A piece of damp earth was taken from the interior of a specimen of this wood, and placed in a room of the palace of Acquasparta, belonging to Duke Cesi. After several months it was found to be completely converted into wood—as seen, not without astonishment, by the aforementioned Lord, and by others who viewed it. And not a single person doubted that earth was the seed and mother of this wood [*la terra è seme e madre di questo legno*].

With twentieth-century hindsight, we can easily understand how Stelluti fell into error and read his story backward. His specimens are ordinary fossil wood, the remains of ancient plants. The actual sequence of transformation runs from real wood, to replacement of wood by percolating minerals (petrifaction), to earth that may either represent weathered and degraded petrified wood, or may just be deposited around or inside the wood by flowing waters. In other words, Stelluti ran the sequence backward in his crucial fourth argument—from formless earth to metallophytes located somewhere between the mineral and vegetable kingdoms!

Moreover, Stelluti's criteria of shaping by overlying sediments (arguments 2 and 3) hold just as well for original wood later distorted and compressed, as for his reversed sequence of metallophytes actively growing within restricted spaces. Delicate parts fossilize only rarely, so the absence of leaves and stems, and the restriction of specimens to trunks, only records the usual pathways of preservation for ancient plants, not Stelluti's naive idea (argument 1) that the tree trunks cannot belong to the vegetable kingdom unless fossilized seeds or roots can also be found. As for the supposedly crucial experiment (argument 5)—well, what can we do with an undocumented three-hundred-year-old verbal report ranking only as hearsay even for Stelluti himself!

Nonetheless, Stelluti's treatise played an important role on the wrong side of the great debate about the nature of fossils—a major issue throughout seventeenth-century science, and not fully resolved until the mid-eighteenth century (see essay 1, about a late defense from 1726). Important authors throughout Europe, from Robert Plot in England (1677), to Olaus Worm in Denmark (1655), reported Stelluti's data as important support for the view that fossils can originate within the mineral kingdom and need not represent the remains of organisms. (Stelluti, by the way, did not confine his arguments to the wood of Acquasparta but made a general extrapolation to the nature and status of all fossils. In a closing argument, depicted on a fateful thirteenth plate of ammonites, Stelluti held that all fossils belong to the mineral kingdom and grow within rocks.)

When we evaluate the logic and rhetoric of Stelluti's arguments, one consistent strategy stands out. Stelluti had finally become a true disciple of Galileo and the primacy of direct empirical observation, viewed as inherently objective. Over and over again, Stelluti states that we must accept his conclusions because he has seen the phenomenon, often several times over many years, with his own eyes.

Stelluti had used this Galilean rhetoric to great advantage before. At the very bottom of his beautiful 1625 engraving of three bees for Pope Urban, Stelluti had added a little Latin note, just under his greatest enlargement of paired bee legs. In a phrase almost identical in form with Galileo's anagram about Saturn, Stelluti wrote: *Franciscus Stellutus Lynceus Fabr*^{is} *Microscopio Observavit*—"the Lynx Francesco Stelluti from [the town of] Fabriano observed [these objects] with a microscope." This time, at least, Stelluti had a leg up on Galileo—for the slow stepper among the Lynxes had made accurate observations, properly interpreted, while Galileo had failed for the much more difficult problem of Saturn. (This note, by the way, may represent the first appearance of the word *microscope* in print. Galileo had called his instrument an *occhiolino,* or "little eye," and his fellow Lynxes had then suggested the modern name.)

But Stelluti's luck had run out with Cesi's wood, when the same claim now buttressed his errors. Consider a sampling, following the order of his text, of Stelluti's appeals to the incontrovertible status of direct observation:

> The generation of this wood, which I have been able to see and observe so many times, does not proceed from seeds . . .

> The material of this wood is nothing other than earth, because I have seen pieces of it [*perche n'ho veduto io pezzi*] with one part made of hard earth and the other of wood.

> Figure 7 shows a drawing of a large oval specimen, which I excavated myself from the earth.

> The outer surface of the other piece appears to be entirely in wood, as is evident to the eye [in the drawing presented by Stelluti].

Stelluti ends his treatise with a flourish in the same mode: he need not write at great length to justify his arguments (and his text only runs to twelve pages), because he has based his work on personal observation:

> And this is all I need to say, with maximal brevity, about this material, which I have been able to see and observe so many times in those places where this new, rare, and marvelous phenomenon of nature originates.

But Stelluti had forgotten the old principle now embodied in a genre of jokes that begin by proclaiming: "I've got some good news, and some bad news." Galileo's empirical method can work wonders. But hardly any faith can be more misleading than an unquestioned personal conviction that the apparent testimony of one's own eyes must provide a purely objective account, scarcely requiring any validation beyond the claim itself. Utterly unbiased observation must rank as a primary myth and shibboleth of science, for we can only see what fits into our mental space, and all description includes interpretation as well as sensory reporting. Moreover, our mental spaces house a complex architecture built of social constraint, historical circumstance, and psychological hope—as well as nature's factuality, seen through a glass darkly.

We can be terribly fooled if we equate apparent sight with necessary physical reality. The great Galileo, the finest scientist of his or any other time, *knew* that Saturn—Stelluti's personal emblem—must be a triple star because he had so observed the farthest planet with good eyes and the best telescope of his day, but through a mind harboring no category for rings around a celestial sphere. Stelluti *knew* that fossil wood must grow from earths of the mineral kingdom because he had made good observations with his eyes and then ran an accurate sequence backward through his mind.

And thus, nature outfoxed the two Lynxes at a crucial claim in their careers—because both men concluded that sight alone should suffice, when genuine solutions demanded insight into mental structures and strictures as well.

As a final irony, Cesi had selected the emblem of Stelluti and Galileo's own society—the lynx—as an exemplar of this richer, dual pathway. The duke of Acquasparta had named his academy for a wild and wily cat, long honored in legend for possessing the sharpest sight among animals. Cesi chose well and subtly—and for a conscious and explicit reason. The maximal acuity of the lynx arose from two paired and complementary virtues—sharpness of vision *and* depth of insight, the outside and the inside, the eye and the mind.

Cesi had taken the emblem for his new society from the title page of Giambattista Della Porta's *Natural Magic* (1589 edition), where the same picture of a lynx stands below the motto: *aspicit et inspicit*—literally meaning "he looks at and he looks into," but metaphorically expressing the twinned ideals of observation and experimentation. Thus, the future fifth Lynx, the living vestige of the old way, had epitomized the richer path gained by combining insight with, if you will, "exsight," or observation. Cesi had stated the ideal in a document of 1616, written to codify the rules and goals of the Lynxes:

> In order to read this great, true and universal book of the world, it is necessary to visit all its parts, and to engage in both observation and experiment in order to reach, by these two good means, an acute and profound contemplation, by first representing things as they are and as they vary, and then by determining how we can change and vary them.

If we decide to embrace the entire universe as our potential domain of knowledge and insight—to use, in other words, the full range of scales revealed by Galileo's two great instruments, the telescope and the microscope (both, by the way, named by his fellow Lynxes)—we had better use all the tools of sensation *and* mentality that a few billion years of evolution have granted to our feeble bodies. The symbol of the lynx, who sees most acutely from the outside, but who also understands most deeply from the inside, remains our best guide. Stelluti himself expressed this richness, this duality, in a wonderfully poetic manner by extolling the lynx in his second major book, his translations of the poet Persius, published in 1630. Cesi had selected the lynx for its legendary acuity of vision, but Stelluti added:

> Not merely of the exterior eyes, but also of the mind, so necessary for the contemplation of nature, as we have taught, and as we practice, in our quest to penetrate into the interior of things, to know the causes and operations of nature . . . just as the lynx, with its superior vision, not only sees what lies outside, but also notes what arises from inside.

3

How the Vulva Stone Became a Brachiopod

WE USUALLY DEPICT THE RENAISSANCE (LITERALLY, THE "rebirth") as a clear, bubbling river of novelty that broke the medieval dam of rigidified scholasticism. But most participants in this great ferment cited the opposite of innovation as their motive. Renaissance thinkers and doers, as the name of their movement implied, looked backward, not forward, as they sought to rediscover and reinstitute the supposed perfection of intellect that Athens and Rome had achieved and a degraded Western culture had forgotten.

I doubt that anyone ever called Francis Bacon (1561–1626) a modest man. Nonetheless, even the muse of ambition must have smiled at such an audacious gesture when this most important British philosopher since the death of William of Ockham in 1347, his chancellor of England (until his fall for financial improprieties), declared "all knowledge" as his "province" and announced that he would write a Great Instauration (defined by *Webster's* as "a restoration after decay, lapse or dilapidation"), both to codify the fruitful rules of reason and to summarize all useful results. As a procedural starting point at the dawn of a movement that would become modern science, Bacon rejected both the scholastic view that equated knowledge with conservation, and the Renaissance reform that sought to recapture a long-lost perfection. Natural knowledge, he proclaimed, must be reconceptualized as a cumulative process of discovery, propelled by processing sensory data about the external world through the reasoning powers of the human brain.

Aristotle's writings on logic had been gathered into a compendium called the *Organon* (or "tool"). Bacon, with his usual flair, entitled the second book of his great instauration the *Novum Organum,* or new tool of reasoning—because the shift to such a different ideal of knowledge as cumulative, and rooted in an increasing understanding of external reality, also demanded that the logic of reasoning itself be reexamined. Bacon therefore began the *Novum Organum* by analyzing impediments to our acquisition of accurate knowledge about the empirical world. Recognizing the existence of such barriers required no novel insight. Aristotle himself had classified the common logical fallacies of human reasoning, while everyone acknowledged the external limits of missing data—stars too far away to study in detail (even with Galileo's newfangled telescope), or cities too long gone to leave any trace of their former existence.

But Bacon presented a brilliant and original analysis by concentrating instead on psychological barriers to knowledge about the natural world. He had, after all, envisioned the study of nature as a funneling of sensory data through mental processors, and he recognized that internal barriers of the second stage could stand as high as the external impediments of sensory limitations. He also understood that the realm of conceptual hangups extended far beyond the cool and abstract logic of Aristotelian reason into our interior world of fears, hopes, needs, feelings, and the structural limits of mental machinery. Bacon therefore developed an incisive metaphor to classify these psychological barriers. He designated such impediments as "idols" and recognized four major categories—*idola tribus* (idols of the tribe), *idola specus* (the cave), *idola fori* (the forum, or marketplace), and *idola theatri* (the theater).

Proceeding from the particular to the general, idols of the cave define the peculiarities of each individual. Some of us panic when we see a mathematical formula; others, for reasons of childhood suppression grafted upon basic temperament, dare not formulate thoughts that might challenge established orders.

Idols of the marketplace, perhaps Bacon's most original concept, designate limits imposed by language—for how can we express, or even formulate, a concept that no words in our language can specify? (In his brilliant story, "Averroës' Search," for example, Jorge Luis Borges—who loved Bacon's work and may well have written this tale to illustrate the idols—imagined the fruitless struggles of the greatest Arabic commentator on Aristotle to understand and translate the master's key concepts of "tragedy" and "comedy," for such notions could not be expressed, or even conceptualized, in Averroës's culture.)

Idols of the theater identify the most obvious category of impediments based on older systems of thought. We will have one hell of a time trying to grasp Darwinism if we maintain absolute and unquestioned fealty to the "old time religion" of Genesis literalism, with an earth no more than a few thousand years old and all organisms created by a deity, ex nihilo and in six days of twenty-four hours. Finally, idols of the tribe—that is, our tribe of *Homo sapiens*—specify those foibles and errors of thinking that transcend the peculiarities of our diverse cultures and reflect the inherited structures and operations of the human brain. Idols of the tribe, in other words, lie deep within the constitution of what we call "human nature" itself.

Bacon emphasized two tribal idols in his examples: our tendency to explain all phenomena throughout the spatial and temporal vastness of the universe by familiar patterns in the only realm we know by direct experience of our own bodies, the domain of objects that live for a few decades and stand a few feet tall; and our propensity to make universal inferences from limited and biased observations, ignoring evident sources of data that do not impact our senses. (Bacon cites the lovely example of a culture convinced that the Sea God saves shipwrecked men who pray for his aid because rescued sailors so testify. A skeptic, presented with this evidence and asked *"whether he did not now confess the divinity of Neptune?* returned this counter-question by way of answer; *yea, but where are they printed, that are drowned?* And there is the same reason of all such like superstitions, as in astrology, dreams, divinations, and the rest.")

In a 1674 translation of the *Great Instauration* (originally written in Latin), Bacon defines the idols in his characteristically pungent prose:

> Idols are the profoundest fallacies of the mind of man. Nor do they deceive in particulars [that is, objects in the external world] . . . but from a corrupt and crookedly-set predisposition of the mind; which doth, as it were, wrest and infect all the anticipations of the understanding. For the mind of man . . . is so far from being like a

> smooth, equal and clear glass, which might sincerely take and reflect the beams of things, according to their true incidence; that it is rather like an enchanted glass, full of superstitions, apparitions, and impostures.

(Gilbert Wats, Bacon's translator, called his subject "a learned man, happily the learned'st that ever lived since the decay of the Grecian and Roman empires, when learning was at a high pitch." Wats also appreciated Bacon's distinctive approach to defining the embryonic field of modern science as accumulating knowledge about the empirical world, obtained by passing sensory data through the biased processing machinery of the brain. Wats described Bacon as "the first that ever joyn'd rational and experimental philosophy in a regular correspondence, which before was either a subtilty of words, or a confusion of matter." He then epitomized Bacon's view in a striking image: "For Truth, as it reflects on us, is a congruent conformity of the intellect to the object . . . when the intellectual globe, and the globe of the world, intermix their beams and irradiations, in a direct line of projection, to the generation of sciences.")

If our primary tribal idol resides in the ancient Greek proverb that "man [meaning all of us] is the measure of all things," then we should not be surprised to find our bodily fingerprints in nearly every assessment, even (or especially) in our words for abstractions—as in the strength of virility (from the Latin *vir,* "adult male"), the immaturity of puerility (from *puer,* "boy"), or the madness of hysteria (originally defined as an inherently feminine disease, from the Greek word for "womb"). Nevertheless, in our proper objection to such sexual stereotyping, we may at least take wry comfort in a general rule of most Indo-European languages (not including English) that assign genders to nouns for inanimate objects. Abstract concepts usually receive feminine designations—so the nobility of (manly) virtue presents herself as *la vertu* in France, while an even more distinctively manly virility also cross-dresses as *la virilité.*

We can, I believe, dig to an even deeper level in identifying tribal idols that probably lie in the evolved and inherited structures of neural wiring—the most basic and inherent substrate of "human nature" itself (if that ill-defined, overused, and much-abused term has any meaning at all). Some properties of human thinking seem so general, so common to all people, that such an evolutionary encoding seems reasonable, at least as a working hypothesis. For example, neurologists have identified areas of the brain apparently dedicated to the perception of faces. (One can easily speculate about the evolutionary value of such a propensity, but we must also recognize that these inherent biases of perception can strongly distort our judgment in other circumstances—Bacon's reason for designating such mental preferences as idols—as when we think we see a face in the random pitting of a large sandstone block on Mars, and then jump to conclusions about alien civilizations. I am not making this story up, by the way; the Martian face remains a staple of "proof" for the UFO and alien abduction crowd.) I suspect that the neural mechanism for facial recognition becomes activated by the abstract pattern of two equal and adjacent circles with a line below—a configuration encountered in many places besides real faces.

In this "deeper" category of tribal idols, I doubt that any rule enjoys wider application, or engenders greater trouble at the same time, than our tendency to order nature by making dichotomous divisions into two opposite groups. (Claude Lévi-Strauss and the French structuralists have based an entire theory of human nature and social history on this premise—and two bits from this corner says they're right, even if a bit overextended in their application.) Thus, we start with a few basic divisions of male versus female and night versus day—and then extend these concrete examples into greater generalities of nature versus culture ("the raw and the cooked" of Lévi-Strauss's book), spirit versus matter (of philosophical dualism), the beautiful versus the sublime (in Burke's theory of aesthetics); and thence (and now often tragically) to ethical belief, anathematization, and, sometimes, warfare and genocide (the good versus the bad, the godly who must prevail versus the diabolical, ripe for burning).

Again, one can speculate about the evolutionary basis of such a strong tendency. In this case, I rather suspect that dichotomization represents some "baggage" from an evolutionary past of much simpler brains built only to reach those quick decisions—fight or flight, sleep or wake, mate or wait—that make all the difference in a Darwinian world. Perhaps we have never been able to transcend the mechanics of a machinery built to generate simple twofold divisions and have had to construct our greater complexities upon such a biased and inadequate substrate—perhaps the most restrictive tribal idol of all.

I devoted the first part of this essay to a general discussion of our mental limitations because this framework, I believe, so well illuminates a particular problem in the history of paleontology that caught my fancy and attention both for unusual intrigue in itself, and for providing such an excellent "test case" of an important general pattern in the growth of scientific knowledge.

Classical authors, particularly Pliny in his *Natural History,* spoke in a limited way about fossils, usually (and correctly) attributing the shells found on mountaintops to a subsequent elevation of land from ancient seabeds. A few medieval authors (particularly Albert the Great in the thirteenth century) added a few comments, while Leonardo da Vinci, in the Leicester Codex (written in the early 1500s), made extensive and brilliant paleontological observations that were, however, not published until the nineteenth century, and therefore had no influence upon the field's later development. Essentially, then, the modern history of paleontology began in the mid-sixteenth century with the publication of two great works by two remarkable scholars: treatises on fossils published in 1546 by the German physician and mining engineer Georgius Agricola, and in 1565 (the year of his death in an epidemic of plague in Zurich) by the Swiss polymath Conrad Gesner.

In the compendium of Latinized folk names then used to identify fossils, most designations noted either a similarity in appearance to some natural or cultural phenomenon, or a presumed and legendary mode of origin. Thus, the flat and circular components of crinoid stems were called *trochites,* or wheel stones; the internal molds of rounded pairs of clamshells were *bucardites* (see accompanying figure, published in 1665), or bull's hearts; well-rounded concretions of the appropriate size were *enorchites,* or testicle stones (and if three were joined together, they became *triorchites,* or "three balls"); sea urchin tests became *brontia* (or thunder stones) because they supposedly fell from the sky in lightning storms.

A prominent group of fossils in this old taxonomy, and a puzzle (as we shall see) to early paleontologists, received the name of *hysteroliths,* also designated, in various vernaculars, as woman stones, womb stones, mother stones, or vulva stones (with the scholarly name derived from the same root as *hysteria,* an example cited earlier in this essay). The basis for this taxonomic consensus stands out in the first drawing of hysteroliths ever published—by the Danish natural historian Olaus Worm in 1665. A prominent median slit on one side

Olaus Worm's 1665 illustration of the internal mold of a clamshell pair. His generation did not know the source of these fossils and called them bucardites, *or "bull's hearts."*

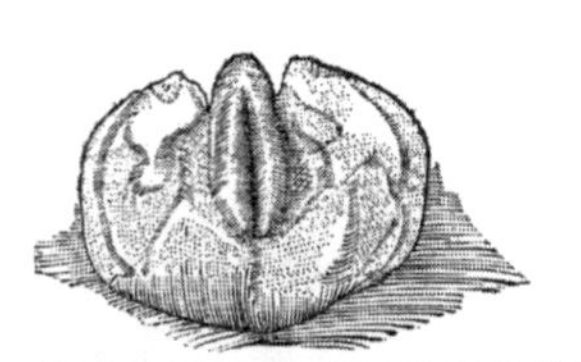

Olaus Worm's original illustration of a hysterolith, *or "womb stone"—actually the internal mold of a brachiopod.*

(sometimes both) of a rounded and flattened object can hardly fail to suggest the anatomical comparison—or to cite Worm's own words, *"quod muliebre pudendum figura exprimat"* (because its form resembles the female genitalia).* Interestingly, as Worm's second figure (to the right) shows, the opposite side of some (but not all) hysteroliths seems to portray a less obvious figure of the male counterparts! The men who wrote the founding treatises of modern paleontology could hardly fail to emphasize such a titillating object (especially in an age that provided few opportunities for approved and legitimate discussion, and illustration, of such intimate subjects).

This essay is not structured as a mystery yarn, so I spoil nothing, while (I hope) enhancing the intended intellectual theme, by providing the solution up front. Hysteroliths are the internal molds of certain brachiopod shells (just as bucardites, discussed and pictured above, are internal molds of certain clamshells). Brachiopods are not closely related to clams, but they also grow shells made of two convex valves that open along a hinge located at one end of the shell, and close by bringing the two valves together along their entire edges. Therefore, if you make an internal mold by pouring plaster of Paris into the closed shell, the resulting object will look roughly like a flattened sphere, with the degree of flattening specified by the convexity of the shell. Highly convex shells can produce nearly spherical molds (as in the fat clamshells that make bucardites). Shells of lower convexity—including most brachiopods and all the groups that make hysteroliths—yield more flattened molds.

Since molds are negative impressions of surrounding shapes, the suggestive parts of hysteroliths record features on the interior of a brachiopod shell in reverse. The slit that suggested a vulva and gave hysteroliths their name marks the negative impression of a raised and narrow linear ridge—called the median septum—that runs right down the middle of many brachiopod shell interiors, effectively dividing the valve in half. (For a clarifying analogy, think of the ridge as a knife and the slit as a cut.) The less pronounced "male" features on the

*Unless otherwise noted, all translations from the literature on hysteroliths are mine from Latin originals.

other side of some hysteroliths record, in positive relief, a cylindrical groove on the shell interior that houses part of the feeding skeleton (detached from the shell itself and rarely fossilized) in some groups of brachiopods.

By the mid-eighteenth century, paleontologists had reached a correct consensus. They knew that hysteroliths were internal molds of brachiopods, and they had even learned which kinds of brachiopods left such impressions upon their molds. They also recognized, of course, that the admittedly striking similarity with human genitalia recorded a sheer, if curious, accident with no causal meaning or connection whatsoever.

We therefore obtain, in the story of hysteroliths, a clean, clear, and lovely example of science operating at its best, by following the canonical definition of its very being and distinctiveness—a procedure dedicated to the sweetest of all goals: the construction of an accurate piece of natural knowledge. This odyssey through two centuries and several interesting stages progresses from the puzzled agnosticism of Agricola's first mention in 1546 toward Linnaeus's unchallenged consensus of 1753. I certainly do not deny the broad outline of this story. Agricola and Gesner possessed few clues for deciding among a wide range of alternatives—from the correct answer that eventually prevailed, to a hypothesis of inorganic origin by plastic forces circulating through rocks, to production by various ancient animals as a meaningful symbol that might even cure or alleviate human ailments of the genital organs. The correct answer may not have fulfilled all human hopes and uses, but hysteroliths really are brachiopod molds, and science supplied the tools for proper resolution.

I do, however, question the usual reading of such genuine scientific progress as a simple exercise in factual accumulation through accurate observation guided by objective principles of reasoning known as the scientific method. In this familiar model, the naïveté of Agricola and Gesner arises from their lack of accurate knowledge, not from any mental failures or barriers. In this sense, these sixteenth-century scholars might well be us in miniature, with the diminution established by what they couldn't know and we have since learned by living several centuries later and enjoying the fruits of advancing scientific understanding. But we should not so diminish these brilliant men and their interesting times. Gesner and Agricola do not rank below us; they only differed from us (while, no doubt, possessing more intrinsic "smarts" than the vast majority of us) in viewing the world from entirely divergent points of view that would be fascinating for us to comprehend.

I particularly appreciate Bacon's metaphor of the idols because this device can lead us toward a better appreciation for the complexities of creative thought, and the unifying similarities between the style we now call science and all other modes of human insight and discovery (while acknowledging, of course, that science presides over distinct subject matter and pursues particular goals in trying to understand the factual character of a "real" external world). Bacon argued that we must filter sensory data about this world through mental processors, and that these internal mechanisms always operate imperfectly because idols gum up the works. Discovery, therefore, arises from a complex intermeshing of these inside and outside components, and not by the accumulated input of facts from the outside world, processed through centuries by the universal and unchanging machinery of internal scientific logic.

Gesner did not use the same criteria for decisions that we employ today, so our differences cannot be attributed to his tiny molehill of reliable facts compared with our mountain. Rather, the idols conspired in him (as they still do in us, but with different resulting blockages) to construct a distinct kind of processing machine. Science prospers as much by retuning, or demolishing and then rebuilding, such mental machinery, as by accumulation of new factual information. Scientists don't simply observe and classify enough fossils until, one day, the status of hysteroliths as brachiopod molds becomes clear; rather, our theories about the nature of reality, and the meaning of explanation itself, must be decomposed and reconstructed before we can build a mental mansion to accommodate such information. And fruitful reconstruction requires, above all, that we acknowledge, examine, and challenge the Baconian idols of our own interior world.

I argued at the beginning of this essay that the Baconian idols could be ordered by degree of generality. In tracing the history of this internal component to solving the problem of hysteroliths, I noted an interesting progression in the release of blockages—from the most pervasive to the most particular idol, as paleontologists homed in upon a solution over two centuries. Perhaps we must first dig the right kind of mine before we can locate any particular nugget of great price.

1. *Idols of the Tribe in the Sixteenth Century.* Gesner and Agricola rediscover Pliny and the three dichotomies.

The hysterolith story begins as long ago as the recorded history of paleontology can venture, and as deeply as one can probe into the most pervasive and general of tribal idols: our propensity to dichotomize. Pliny the Elder, the great Roman statesman and natural historian who died with his boots on in the eruption of Mount Vesuvius in A.D. 79, wrote a compendium about the natural world that survived as legions of hand copies made by monks and other scholars for more than a millennium before Gutenberg, and then became one of the most widely published books in the first decades of printing. (In the trade, books printed before 1500 are called *incunabulae,* or "from the cradle.")

Agricola and Gesner, as Renaissance scholars committed to the recovery of ancient wisdom, sought above all to assign their specimens (and vernacular names) to forms and categories mentioned by Pliny in his *Natural History.* In an alphabetical list of rocks, minerals, and fossils, featured in the thirty-seventh and last book of his great treatise, Pliny included a notable one-liner under letter *D: "Diphyes duplex, candida ac nigra, mas ac femina"*—having the character of both sexes, white and black, male and female.

Pliny's treatise contained no pictures, so we can hardly know what object he had meant to designate with this sparse description. But on the theme of tribal idols, I am fascinated that the first mention of a possible hysterolith features two of the most general impediments in this category: our tendency to read nature at all scales in terms of immediately familiar objects, particularly the human body, and our propensity for classification by dichotomy. Pliny, in fact, explicitly cites two of the most fundamental dichotomies in his single line: male and female, and white and black. (Later commentators assumed that Pliny's *diphyes* referred to stones that looked male on one side and female on the other—hence their identification with hysteroliths.)

Moreover, we should also note the implicit inclusion of a third great dichotomy—top and bottom—in Pliny's definition, for hysteroliths are composed of two distinct and opposite halves, a stunning representation, literally set in stone, of our strongest mental idol, expressed geometrically. Moreover, all three dichotomies carry great emotional weight both in their archetypal ideological status and in their embodiment of conventional rankings (by worth and moral status) in a hierarchical and xenophobic society: male, white, and top versus female, black, and bottom. A modern perspective that we view as far more valid, in both factual and moral terms, can only cause us to shiver when we grasp the full implication of such a multiply dichotomized classification.

In his *De natura fossilium* of 1546, the first published treatise on paleontology (although the term *fossil* then designated any object found in the ground—a broad usage consistent with its status as past participle of the Latin verb *fodere,* "to dig up"—so this work treated all varieties of rocks, minerals, and the remains of organisms now exclusively called fossils), Georgius Agricola unearthed Pliny's one-liner, probably for the first time since antiquity, and applied the name *diphyes* to some fossils found near the fortress of

Ehrenbreitstein. A generation later, in his *De rerum fossilium* (On fossil objects) of 1565, Conrad Gesner first connected Pliny's name and Agricola's objects with the folk designations and Latin moniker—*hysterolith*—that would then denote this group of fossils until their status as brachiopod molds became clear two hundred years later.

Sixteenth-century paleontology proceeded no further with hysteroliths, but we should not undervalue the achievements of Agricola and Gesner in terms of their own expressed aims. As men of the Renaissance, they wished to unite modern observations to classical wisdom—and the application of Pliny's forgotten and undocumented name to a clear category of appropriate objects seemed, to them, an achievement worth celebrating.

Moreover, when we note Gesner's placement of hysteroliths within his general taxonomy of fossils, we can peek through this window into the different intellectual domain of sixteenth-century explanation, and also begin to appreciate the general shifts in worldview that would have to occur before hysteroliths could be recognized as brachiopod molds. Gesner established fifteen categories, mostly based on presumed resemblances to more familiar parts of nature, and descending in a line of worth from the most heavenly, regular, and ethereal to the roughest and lowest. The first category included geometric forms (fossils of circular or spherical shape, for example); the second brought together all fossils that recalled heavenly bodies (including star-shaped elements of crinoid columns); while the third held stones that supposedly fell from the sky. At the other end, the disparaged fossils of class 15 resembled insects and serpents. Gesner placed hysteroliths into category 12, not at the bottom but not very near the honored pinnacle either, for "those that have some resemblance to men or quadrupedal animals, or are found within them." As his first illustration in category 12, Gesner drew a specimen of native silver that looked like a mat of human hair.

2. *Idols of the Theater in the Seventeenth Century.* Animal or mineral; useful symbol or meaningless accident?

If classic tribal idols played a founding role in setting the very name and definition of hysteroliths—their designation for some particularly salient features of female anatomy, and their description, by Pliny himself, in terms of three basic dichotomies that build the framework of our mental architecture—then some equally important theatrical idols (that is, constraints imposed by older, traditional systems of thought) underlay the major debate about the origin and meaning of hysteroliths that only began with seventeenth-century paleontology, but then pervaded the century: what are fossils?

The view of mechanism and causality that we call modern science answers this question without any ambiguity: fossils look like organisms in all their complex details, and we find them in rocks that formed in environments where modern relatives of these creatures now live. Therefore, fossils are remains of ancient organisms. This commonsense view had developed in ancient Greek times, and always held status as an available hypothesis. But the domain of seventeenth-century thought, the world that Bacon challenged and that modern science would eventually supplant, included other alternatives that may seem risible today, but that made eminent sense under other constructions of natural reality.

Bacon called these alternative worldviews idols of the theater, or impediments set by unfruitful systems of thought. Among the theatrical idols of seventeenth-century life, none held higher status among students of fossils than the Neoplatonic construction of nature as a static and eternal set of symbolic correspondences that reveal the wisdom and harmonious order of creating forces, and that humans might exploit for medical and spiritual benefit. A network of formal relationships (not direct causal connections, but symbolic resemblances in essential properties) pervaded the three kingdoms of nature—animal, vegetable, and mineral—placing any object of one kingdom into meaningful correspondence with counterparts in each of the other two. If we could specify and understand this network, we might hold the key to nature's construction, meaning, and utility.

Within this Neoplatonic framework, a close resemblance between a petrified "fish" enclosed within a rock and a trout swimming in a stream does not identify the stony version as a genuine former organism of flesh and blood, but suggests instead that plastic forces within the mineral kingdom can generate this archetypal form within a rock just as animate forces of another kingdom can grow a trout from an egg. Similarly, if various stones look like parts of the human body, then perhaps we can identify the mineral forces that resonate in maximal sympathy with the sources of our own animate being. Moreover, according to a theory of medicine now regarded as kooky and magical, but then perfectly respectable in a Neoplatonic framework, if we could identify the vegetable and mineral counterparts of human organs, then we might derive cures by potentiating our ailing animal versions with the proper sympathies of other realms, for every part in the microcosm of the human body must vibrate in harmony with a designated counterpart in the macrocosm of the earth, the central body of the universe. If the ingested powder of a pulverized "foot stone" could soothe the pains of gout, then hysteroliths might also alleviate sexual disorders.

The availability of this alternative view, based on the theatrical idol of Neoplatonism, set the primary context for seventeenth-century discussions about hysteroliths. Scholars could hardly ask: "what animal makes this shape as its mold?" when they remained stymied by the logically prior, and much more important, question: "are hysteroliths remains of organisms or products of the mineral kingdom?" This framework then implied another primary question—also posed as a dichotomy (and thus illustrating the continuing intrusion of tribal idols as well)—among supporters of an inorganic origin for hysteroliths: if vulva stones originate within the mineral kingdom, does their resemblance to female genitalia reveal a deep harmony in nature, or does the similarity arise by accident and therefore embody no meaning, a mode of origin that scholars of the time called *lusus naturae,* a game or sport of nature?

To cite examples of these two views from an unfamiliar age, Olaus Worm spoke of a meaningful correspondence in 1665, in the textual commentary to his first pictorial representation of hysteroliths—although he attributed the opinion to someone else, perhaps to allay any suspicion of partisanship:

> These specimens were sent to me by the most learned Dr. J. D. Horst, the archiater [chief physician] to the most illustrious Landgrave of Darmstadt. . . . Dr. Horst states the following about the strength of these objects: these stones are, without doubt, useful in treating any loosening or constriction of the womb in females. And I think it not silly to believe, especially given the form of these objects [I assume that Dr. Horst refers here to hysteroliths that resemble female parts on one side and male features on the other] that, if worn suspended around the neck, they will give strength to people experiencing problems with virility, either through fear or weakness, thus promoting the interests of Venus in both sexes [*Venerem in utroque sexu promovere*].

But Worm's enthusiasm did not generate universal approbation among scholars who considered an origin for hysteroliths within the mineral kingdom. Anselm de Boot, in the 1644 French translation of his popular compendium on fossils (in the broad sense of "anything found underground"), writes laconically: *"Elles n'ont aucune usage que je sçache"* (they have no use that I know).

By the time that J. C. Kundmann—writing in vernacular German and living in Bratislava, relatively isolated from the "happening" centers of European intellectual life—presented the last serious defense for the inorganic theory of fossils in 1737, the comfortable rug of Neoplatonism had already been snatched away by time. (The great Jesuit scholar Athanasius Kircher had written the last major defense of Neoplatonism in paleontology in 1664, in his *Mundus subterraneus,* or *Underground World.*) Kundmann therefore enjoyed little intellectual maneuvering room beyond a statement that the resemblances to female genitalia could only be accidental—for, after all, he argued, a slit in a round rock can arise by many mechanical routes. In a long chapter devoted to hysteroliths, Kundmann allowed that hysteroliths might be internal molds of shells, and even admitted that some examples described by others might be so formed. But he defended an inorganic origin for his own specimens because he found no evidence of any surrounding shell material: "an excellent argument that these stones have nothing to do with clamshells, and must be considered as *Lapides sui generis*" (figured stones that arise by their own generation—a "signature phrase" used by supporters of an inorganic origin for fossils).

3. *Idols of the Marketplace in the Eighteenth Century.* Reordering the language of classification to potentiate the correct answer.

As stated above, the inorganic theory lost its best potential rationale when the late-seventeenth-century triumph of modern scientific styles of thinking (the movement of Newton's generation that historians of science call "*the* scientific revolution") doomed Neoplatonism as a mode of acceptable explanation. In this new eighteenth-century context, with the organic theory of fossils victorious by default, a clear path should have opened toward a proper interpretation of hysteroliths.

But Bacon, in his most insightful argument of all, had recognized that even when old theories (idols of the theater) die, and when deep biases of human nature (idols of the tribe) can be recognized and discounted, we may still be impeded by the language we use and the pictures we draw—idols of the marketplace, where people gather to converse. Indeed, in eighteenth-century paleontology, the accepted language of description, and the traditional schemes of classification (often passively passed on from a former Neoplatonic heritage without recognition of the biases thus imposed) established major and final barriers to solving the old problem of the nature of hysteroliths.

At the most fundamental level, remains of organisms had finally been separated as a category from other "things in rocks" that happened to look like parts or products of the animal and vegetable kingdom. But this newly restricted category commanded no name of its own, for the word *fossil* still covered everything found underground (and would continue to do so until the early nineteenth century). Scholars proposed various solutions—for example, calling organic remains "extraneous fossils" because they entered the mineral kingdom from other realms, while designating rocks and minerals as "intrinsic fossils"—but no consensus developed during the eighteenth century. In 1804, the British amateur paleontologist James Parkinson (a physician by day job, and the man who gave his name to Parkinson's disease), recognizing the power of Bacon's idols of the marketplace and deploring this linguistic impediment, argued that classes without names could not be properly explained or even conceptualized:

> But when the discovery was made, that most of these figured stones were remains of subjects of the vegetable and animal kingdoms, these modes of expression were found insufficient; and, whilst endeavoring to find appropriate terms, a considerable difficulty arose; language not possessing a sign to represent that idea, which the mind of man had not till now conceived.

The retention of older categories of classification for subgroups of fossils imposed an even greater linguistic restriction. For example, so long as some paleontologists continued to use such general categories as *lapides idiomorphoi* ("figured stones"), true organic remains would never be properly distinguished from accidental resemblances (a concretion recalling an owl's head, an agate displaying in its color banding a rough picture of Jesus dying on the cross, to cite two actual cases widely discussed by eighteenth-century scholars). And absent such a separation, and a clear assignment of hysteroliths to the animal kingdom, why should anyone favor the hypothesis of brachiopod molds, when the very name *vulva stone* suggested a primary residence in the category of accidents—

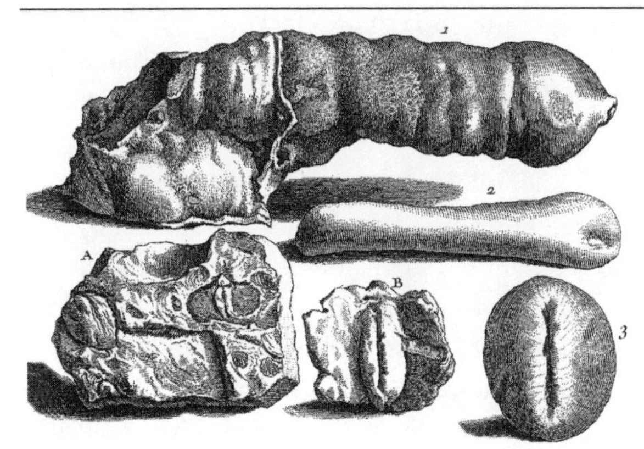

A 1755 illustration of hysteroliths on the same plate as a stalactite that accidentally resembles a penis.

for no one had ever argued that hysteroliths could be actual fossilized remains of detached parts of female bodies!

As a pictorial example, consider the taxonomic placement of hysteroliths in a 1755 treatise by the French natural historian Dezallier d'Argenville. He draws his true hysterolith (Figure A in the accompanying illustration) right next to slits in rocks that arose for other reasons (B and 3) and, more importantly, right under a stalactite that happens to look like a penis with two appended testicles. Now we know that the stalactite originated from dripping calcite in a cave, so we recognize this unusual resemblance as accidental. But if hysteroliths really belong in the same taxonomic category, why should we regard them as formed in any fundamentally different way?

When these idols of the marketplace finally receded, and hysteroliths joined other remains of plants and animals in an exclusive category of organic remains—and when the name *hysterolith* itself, as a vestige of a different view that emphasized accidental resemblance over actual mode of origin, finally faded from use—these objects could finally be seen and judged in a proper light for potential resolution.

Even then, the correct consensus did not burst forth all at once, but developed more slowly, and through several stages, as scientists, now and finally on the right track, moved toward a solution by answering a series of questions—all dichotomously framed once again—that eventually reached the correct solution by successive restriction and convergence. First, are hysteroliths molds of an organism, or actual petrified parts or wholes? Some proposals in the second category now seem far-fetched—for example, Lang in 1708 on hysteroliths as fossilized sea anemones of the coral phylum (colonies of some species do grow with a large slit on top), or Barrère in 1746 on *cunnulites* (as he called them, with an obvious etymology not requiring further explanation on my part) as the end pieces of long bones (femora and humeri) in juvenile vertebrates, before these termini fuse with the main shaft in adulthood. But at least paleontologists now operated within a consensus that recognized hysteroliths as remains of organisms.

Second, are hysteroliths the molds of plants or animals, with nuts and clams as major contenders in each kingdom—and with a quick and decisive victory for the animal kingdom in this case. Third, and finally, are hysteroliths the internal molds of clams or brachiopods—a debate that now, at the very end of the story, really could be solved by something close to pure observation, for consensus had finally been reached on what questions to ask and how they might be answered. Once enough interiors of brachiopod shells had been examined—

not so easy because almost all brachiopod fossils expose the outside of the shell, while few living brachiopods had been observed (for these animals live mostly in deep waters, or in dark crevices within shallower seas)—the answer could not be long delayed.

We may close this happy tale of virtue (for both sexes) and knowledge triumphant by citing words and pictures from two of the most celebrated intellectuals of the eighteenth century. In 1773, Elie Bertrand published a classification of fossils commissioned by Voltaire himself as a guide for arranging collections. His preface, addressed to Voltaire, defends the criterion of mode of origin as the basis for a proper classification—a good epitome for the central theme of this essay. Turning specifically to hysteroliths, Bertrand advises his patron:

> There is almost no shell, which does not form internal molds, sometimes with the shell still covering the mold, but often with only the mold preserved, though this mold will display all the interior marks of the shell that has been destroyed. This is the situation encountered in hysteroliths, for example, whose origin has been debated for so long. They are the internal molds of . . . terebratulids [a group of brachiopods]. (My translation from Bertrand's French.)

But if a good picture can balance thousands of words, consider the elegant statement made by Linnaeus himself in the catalog of Count Tessin's collection that he published in 1753. The hysteroliths (Figure 2, A–D), depicted with both their male and female resemblances, stand next to other brachiopod molds that do not resemble human genitalia (Figure 1, A and B)—thus establishing the overall category by zoological affinity rather than by external appearance. In Figures 3 to 7, following, Linnaeus seals his case by drawing the fossilized shells of related brachiopods. Two pictures to guide and establish a transition—from the lost and superseded world of Dezallier d'Argenville's theory of meaning by accidental resemblance to distant objects of other domains, to Linnaeus's modern classification by physical origin rather than superficial appearance.

Bacon's idols can help or harm us along these difficult and perilous paths to accurate factual knowledge of nature. Idols of the tribe may lie deep within the structure of human nature, but we should also thank our evolutionary constitution for another ineradicable trait of mind that will keep us going and questioning until we break through these constraining idols—our drive to ask and to know. We cannot look at the sky and not wonder why we see blue. We

In 1753, Linnaeus recognized hysteroliths as brachiopod molds and illustrated them with other brachiopods that do not mimic female genitalia.

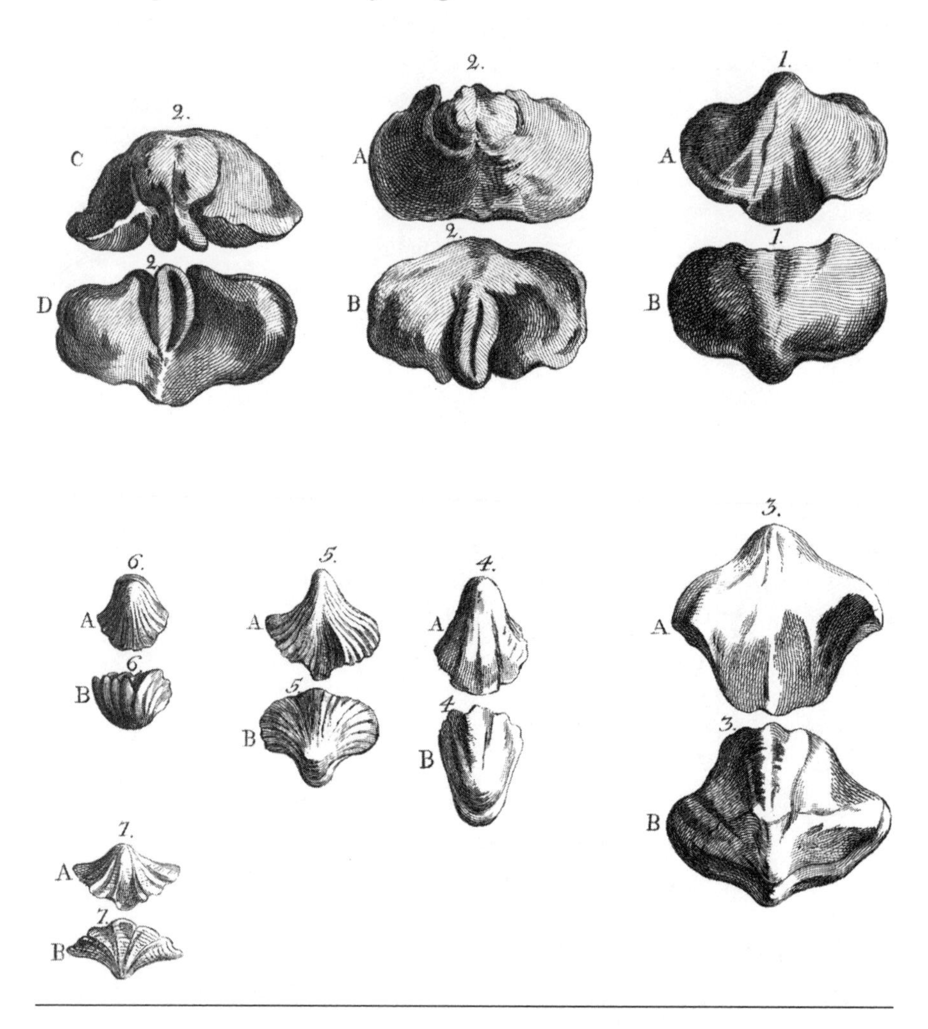

cannot observe that lightning kills good and bad people alike without demanding to know why. The first question has an answer; the second does not, at least in the terms that prompt our inquiry. But we cannot stop asking.

Let me close by tying the two parts of this essay together with a story that unites Bacon (the anchor of the first part) with Pliny (the progenitor of the second part) in their common commitment to this liberating compulsion to ask and know. Pliny died because he could not forgo a unique opportunity to learn something about the natural world—as he sailed too close to the noxious fumes of Vesuvius when he wished to observe a volcanic eruption more closely. Bacon died, albeit less dramatically, in the same noble cause and manner when he devised an experiment one cold day to determine whether snow could retard putrefaction. He stopped his carriage, bought a hen from a poultryman, and stuffed it with snow. The experiment worked, but the doctor died (not the patient this time, for the hen had expired before the procedure began!), as Bacon developed a cold that progressed to bronchitis, pneumonia, and death. He wrote a touching last letter (also quoted in a footnote to chapter 7) expressing a last wish for an explicit connection with Pliny: "I was likely to have the fortune of Caius Plinius the elder, who lost his life by trying an experiment about the burning of the mountain Vesuvius: for I was also desirous to try an experiment or two, touching on the conversion and induration of bodies. As for the experiment itself, it succeeded excellently well, but . . ."

Tribal idols may surround us, but our obsessively stubborn tribal need to ask and know can also push us through, as we follow Jesus' dictum that the truth will make us free. But we must also remember that Jesus then declined to answer Pilate's question: "what is truth?" Perhaps he understood that the idols conspire within us to convert this apparently simple inquiry into the most difficult question of all. But then, Jesus also knew, from the core of his being (in the conventional Christian interpretation), that human nature features an indivisible mixture of earthy constraint and (metaphorically at least) heavenly possibilities for liberation by knowledge—a paradox that virtually defines both the fascination and the frustration of human existence. We needed two hundred years of debate and discovery to turn a vulva stone into a brachiopod; but the same process has also stretched our understanding out to distant galaxies and back to the big bang.

II

Present at the Creation

How France's Three Finest Scientists Established Natural History in an Age of Revolution

4

Inventing Natural History in Style

Buffon's Style and Substance

An average nobleman in eighteenth-century France, including his wig, did not match the modern American mean in height. Nonetheless, at a shade under five feet five, Georges-Louis Leclerc, comte de Buffon, struck his own countrymen as short of stature. Yet he bestrode his world like a colossus. When he died, in 1788 at age eighty, his autopsy, performed by his own prior mandate, yielded fifty-seven bladder stones and a brain "of slightly larger size than that of ordinary [men]." Fourteen liveried horses, nineteen servants, sixty clerics, and a choir of thirty-six voices led his burial procession. The *Mercure* reported:

> His funeral rites were of a splendor rarely accorded to power, opulence, dignity. . . . Such was the influence of this famous name that 20,000 spectators waited for this sad procession, in the streets, in the windows, and almost on the rooftops, with that curiosity that the people reserve for princes.

Buffon lived to see the first thirty-six volumes of his monumental *Histoire naturelle* (written with several collaborators, but under his firm and meticulous direction at all times); the remaining eight tomes were published after his death. No other eighteenth-century biologist enjoyed a wider readership or greater influence (with the possible exception of his archrival Linnaeus). Yet outside professional circles, we hardly recognize Buffon's name today. His one "standard" quotation—*"le style c'est l'homme même"* (style is the man himself)—comes from his inaugural address following his election as one of the "forty immortals" of the Académie Française, and not from his scientific publications. (See Jacques Roger's remarkable book, *Buffon's Life and Works,* translated by Sarah Lucille Bonnefoi, Cornell University Press, 1997.)

We must not equate the fading of a name through time with the extinction of a person's influence. In so doing, we propagate one of the many errors inspired by our generation's fundamental confusion of celebrity with stature. I will argue that, under certain definite circumstances—all exemplified in Buffon's life and career—a loss of personal recognition through time actually measures the spread of impact, as innovations become so "obvious" and "automatic" that we lose memory of sources and assign their status to elementary logic from time immemorial. (I do not, of course, challenge the truism that most fadings record the passage of a truly transient reason for celebrity; Linda Tripp and Tonya Harding come immediately to my mind, but surely not to the consciousness of any future grandchildren.)

Two prerequisites of intellectual fame have been well recognized: the gift of extraordinary intelligence, and the luck of unusual circumstances (time, social class, and so forth). I believe that a third factor—temperament—has not been given its equal due. At least in my limited observation of our currently depleted world, the temperamental factor seems least variable of all. Among people I have met, the few whom I would term "great" all share a kind of unquestioned, fierce dedication; an utter lack of doubt about the value of their activities (or at least an internal impulse that drives through any such angst); and above all, a capacity to work (or at least to be mentally alert for unexpected insights) at every available moment of every day in their lives. I have known other people of equal or greater intellectual talent who succumbed to mental illness, self-doubt, or plain old-fashioned laziness.

This maniacal single-mindedness, this fire in the belly, this stance that sets the literal meaning of *enthusiasm* ("the intake of God"), defines a small group of people who genuinely merit the cliché of "larger than life"—for they seem to live on another plane than we petty men who peep about under their huge legs. This mania bears no particular relationship to the external manifestation known as charisma. Some people in this category bring others along by exuding their zest; others may be glumly silent or actively dyspeptic toward the rest of the world. This temperament establishes an internal contract between you and your muse.

Buffon, all five feet and a bit of him, surely stood larger than life in this crucial sense. He established a rhythm of work in early adulthood and never deviated until his brief and final illness. Every spring, he traveled to his estate at Montbard in Burgundy, where he wrote *Histoire naturelle* and acted out the full life of a tough but fair seigneur and a restless entrepreneur (working to extend his agricultural projects, or building forges to smelt the local iron ore). Every fall he returned to Paris, where he dealt and cajoled to transform the Royal Botanical Garden (which he directed) into the finest general natural history museum in the world—a position certainly achieved by the following generation (and arguably still maintained today) when the successor to Buffon's expansion, the Muséum d'Histoire Naturelle, featured the world's three greatest naturalists as curators: Jean-Baptiste Lamarck, Georges Cuvier, and Etienne Geoffroy Saint-Hilaire.

Buffon worked at least fourteen hours every day. (He refused to alter any detail of this regimen, even in his last years when bladder stones, and various other maladies of old age, made travel so painful.) Jacques Roger describes the drill: "Those who worked with him or were under his orders had to adapt to his lifestyle. And everywhere, the same rule was in force: do not waste time." Buffon himself—in a passage that gives a good taste of the famous style (equal to the man himself!) of *Histoire naturelle*—attacked the Stoics with his personal formula for a life of continual enjoyment and action. If we accede to the stoical view, Buffon warned:

> Then let us say . . . that it is sweeter to vegetate than to live, to want nothing rather than satisfy one's appetite, to sleep a listless sleep rather than open one's eyes to see and to sense; let us consent to leave our soul in numbness, our mind in darkness, never to use either the one or the other, to put ourselves below the animals, and finally to be only masses of brute matter attached to the earth.

As for the other two prerequisites, the necessary brilliance shines forth in Buffon's work and needs no further comment. But Buffon's circumstances should have precluded his achievements (if temperament and brilliance had not pushed him through). As the son of a successful bourgeois family in Burgundy, he was not badly born (he received his later title of count from King Louis XV, and for his own efforts). But science, as a career, scarcely existed in his time—and non-Parisian nonnobility had little access to the few available opportunities. Buffon got a good education at a Jesuit *lycée* in Dijon, and he showed particular early talent for a field quite different from the later source of his triumph: mathematics. He wrote an important treatise on probability, translated Newton's *Fluxions* into French (from an English version of the Latin original), and applied his quantitative skills to important studies on the strength of timber grown on his own estate. He then worked through this botanical door to his eventual post as director of the king's gardens in Paris. The rest, as they say, is (natural) history.

Thirty-six volumes of *Natural History* appeared under Buffon's explicit authorship during his lifetime—one of the most comprehensive and monumental efforts ever made by one man (with a little help from his friends, of course) in science or literature. He intended to cover the entire range of natural objects in all three conventional kingdoms of animal, vegetable, and mineral. In truth, for he started at the traditional top and worked down, he never got to invertebrates or plants (or rather, he bypassed these "lower" manifestations of organic matter to write several volumes, late in life, on what he called "my dear minerals"). Moreover, despite plans and sketches, his own work on vertebrates didn't proceed "below" mammals and birds—and his colleague Lacépède published the last eight volumes (for a total of forty-four in the complete first edition) on reptiles and fishes (including whales) after Buffon's death.

Buffon treated all the great subjects of natural history in their full generality—from geology, to the origin of life, to embryology, physiology, biogeography, functional anatomy, and systematics. He regarded humans as a species of animal with unique properties, and therefore also covered most of anthropology, sociology, and cultural history as well. The general and theoretical articles of *Natural History* inspired endless and passionate debate—and made him a rarity in the history of literature: a man who became wealthy by his wits. (Inheritance and patronage didn't hurt either, but Buffon's volumes were best-sellers.) All sectors of French intellectual life, from the Encyclopedists to the Theological Faculty of the Sorbonne, took up his themes with gusto (agreeing with some and lambasting others, for Buffon's work was too multifarious, and too nuanced, for anyone's outright approbation or dismissal). He fought and made up with Voltaire, Rousseau, and nearly anyone who mattered in the closing years of the *ancien régime.*

But these general articles do not form the heart of *Natural History*. Rather, more than twenty volumes present long, beautifully crafted, descriptively detailed, and passionately opinionated treatises on mammals, birds, and minerals—with each species or kind granted its own chapter. These pieces, illustrated with engravings that became "standard," largely through endless pirating in later works by other authors, remain as charming (and often infuriating) as ever. As an example, consider Buffon's summary comments on his least favorite mammal, the sloth. (I imagine that Buffon, living at his own frenetic level, had even less patience with these slow creatures than those of us who operate at an ordinary human pace can muster):

> Whereas nature appears to us live, vibrant, and enthusiastic in producing monkeys; so is she slow, constrained and restricted in sloths. And we must speak more of wretchedness than laziness—more of default, deprivation, and defect in their constitution: no incisor or canine teeth, small and covered eyes, a thick and heavy jaw, flattened hair that looks like dried grass . . . legs too short, badly turned, and badly terminated . . . no separately movable digits, but two or three excessively long nails. . . . Slowness, stupidity, neglect of its own body, and even habitual sadness, result from this bizarre and neglected conformation. No weapons for attack or defense; no means of security; no resource of safety in escape; confined, not to a country, but to a tiny mote of earth—the tree under which it was born; a prisoner in the middle of great space . . . everything about them announces their misery; they are imperfect productions made by nature, which, scarcely having the ability to exist at all, can only persist for a while, and shall then be effaced from the list of beings. . . . These sloths are the lowest term of existence in the order of animals with flesh and blood; one more defect would have made their existence impossible. (My translation.)

I cannot begin to make a useful summary of the theoretical content of *Histoire naturelle,* if only because Buffon follows Bacon's lead in taking all (at least natural) knowledge for his province, and because Buffon's views do not always maintain full consistency either within or between sections. But short comments on three central subjects may provide some flavor of Buffon's approach to life, and his most important contributions to later research:

1. *Classification.* Carolus Linnaeus, Buffon's Swedish rival and close contemporary (both born in 1707, with Linnaeus dying ten years earlier than Buffon in 1778), developed the system of nomenclature that we continue to use today. Linnaeus prevailed because the formal rules of his system work well in practical terms, and also because his nested and hierarchical scheme of smaller-within-larger categories (species like dogs, within families like canids, within orders like carnivores, within classes like mammals, within phyla like vertebrates) could be slotted into a genealogical interpretation—the arborescent tree of life, with twigs on branches, on boughs, on trunks—that the discovery of evolution would soon impose upon any formal system of naming.

Buffon, on the other hand, sought to encompass all the overt complexity of organisms into a nonhierarchical system that recognized differing relationships for various properties (bats more like mammals in anatomy, more like birds in function). But this alternative model of a network with multiple linkages, rather than a strict hierarchy of inclusion, fails (in the admittedly retrospective light of evolution) to separate the superficial similarity of independent adaptation (wings of bats and birds) from the deep genealogical linkages of physical continuity through the ages (hair and live birth of bats and bears). Buffon's noble vision of equal treatment for all aspects of a species's life—placing ecology, function, and behavior at par with traditional anatomy—foundered on a false theory about the nature of relationships.

2. *Biogeography.* Previous naturalists, if they considered the question at all, generally envisaged a single center of creation for all animals, followed by spreading throughout the globe (a theory obviously consistent with the scenario of the biblical deluge, although not necessarily so inspired or defended). Buffon, on the other hand, recognizing that each species seemed to possess unique adaptations for its own region, argued for origination in appropriate places all over the globe, with only limited subsequent opportunities for migration—a more fruitful idea that founded the modern science of biogeography.

Buffon's notion of adaptation to local conditions directly inspired an important line of research in early American natural history. He argued that American mammals must be smaller than their Old World counterparts (rhino, giraffe, and tiger larger than tapir, llama, and jaguar, for example) because "the heat was in general much less in this part of the world, and the humidity much greater." American naturalists, Thomas Jefferson in particular, became annoyed at this charge of lesser stature for their New World, and sought vigorously to refute Buffon. This strong feeling led Jefferson to his most embarrassing error, when he misidentified the claw of a large fossil ground sloth (ironically, given Buffon's judgment of these creatures) as belonging to a giant lion that would have surpassed all European relatives in bulk. In correcting Jefferson's error, Georges Cuvier diplomatically named this new genus of sloths *Megalonyx jeffersoni*.

3. *The evolution and nature of species.* Most previous systems sought to define these basic units (for groups of organisms) in terms of unique structural features shared by all members and absent from all organisms in other species—an essentialist criterion doomed to failure in our actual world of shadings and exceptions. Buffon, on the other hand, sought a definition rooted in the status and behavior of groups in nature. He therefore held that the ability to interbreed with other members of the species, and to produce healthy and fertile offspring, must become the primary criterion for delimiting the boundaries of natural groups. In so doing, he laid the groundwork for modern notions of the interacting population as nature's basic entity, thus refuting the old Platonic alternative of searching for essential defining features to link any accidental configuration of actual matter (that is, a real organism) to the idealized *eidos* or archetype of its permanent species.

The venerable (and pernicious) tradition of defining past worthies by their supposed anticipation of modern views has misled many commentators into elevating this ecological definition of species, with its rejection of fixed Platonic archetypes, into a prototypical theory of evolution—thus making Buffon the worthy precursor of Darwin on a rectilinear path to truth. But such selective raiding parties from present knowledge into coherent, but fundamentally different, systems of past thought can only derail any effort to grasp the history of ideas as a fascinating panoply of changing worldviews, each fully developed in itself and worthy of our respect and understanding, despite the inevitability (if science has any value at all) of subsequent reformulations that will bring us closer to nature's modes and causes.

Buffon was not, and could not have been, an evolutionist in any modern sense (although *Histoire naturelle,* like the Bible, is so long and various that almost any position can be defended by partial quotation out of context). His system did allow for limited change within original species defined by their capacity for interbreeding. Buffon described these minor alterations as "degenerations" induced by changing climates. (In using the term *degeneration,* he did not invoke our modern meaning of "deterioration"—for such changes usually improved a species's adaptation to local environments—but rather a departure from the "interior mold" or internal guardian of a species's identity in development.)

Buffon's complex and confusing notion of the *moule intérieur* (or "interior mold") underlay his basic theories both of embryology and of life's history through time. He accepted Aristotle's distinction between the controlling *form* of a species and the actual *matter* that builds any particular organism. He rejected Plato's notion of an external and eternal form, accepting Aristotle's alternative view of form as an attribute that shapes labile matter *from within.* For Buffon, the *moule intérieur* acts as the guardian of form and cannot be as labile as matter itself (or very plastic at all), lest general order disappear in nature (an unthinkable notion for an Enlightenment rationalist like Buffon), with each creature becoming no more than a glop of putty shaped only by the accidents of immediately surrounding conditions. For Buffon, a full theory of evolution would have destroyed the rational, albeit complex, order that he had pledged to define in his inimitable style.

BUFFON'S REPUTATION

If Buffon so shaped the science of his day, why did his name not survive as well as the imprint of his ideas? We can identify and distinguish several reasons, each relevant to the coordinating issue that I raised at the outset of this essay: the scaling of reputation with time, and the frequent failure of enduring fame to match continuing influence.

The sound bite does not just represent an invention of modern media in a restless age that has forgotten history. People have always needed simple labels to remember the reasons and meanings of events that shape our past. Unless such a distinctive label can be attached to a person's accomplishments, he will probably fade from sight. The major worthies and icons of the history of science all wear such labels (at least for popular recognition)—Copernicus for a new arrangement of the solar system, Newton for gravity, Darwin for evolution, Einstein for relativity (even if most of us can't define the concept very well). The principle extends beyond intellectual history; for everyone needs such a hook—Pandora her box, Lady Godiva her hair, Mark McGwire his bat. The generality also features a dark side, as good people with strong and consistent accomplishments become inevitably identified by an unforgettable and highly public moment of ultimate chagrin—Bill Buckner for a ball that bounced between his legs; another Bill for something else between his legs.*

*I wrote this essay in the summer of 1998, *in medio Monicae anni,* just before a presidential impeachment.

Buffon had a passion for order, but he developed no central theory that could be defined by a memorable phrase or concept. He wrote volumes of incomparable prose and propagated ideas, sometimes quite radical, about all major subjects in natural history. But no central thread unites his system. Moreover, Buffon may have been just a bit too worldly, just a tad too practical, ever to develop a transforming worldview clear and coherent enough (like Darwin's natural selection) both to attach distinctively to his person, and to apply consistently to a natural world strongly altered thereby.

In his uncomfortable duality of being both larger than life, and also so much *in* the life of his own society, Buffon often had to juggle and feint, to smooth over or to hide under, so that his readers or anyone in power, from priest to patron to Parisian pol, would not dismiss him as too far outside the sensibilities of his surrounding world. Buffon possessed a radical streak, the stubborn independence of all great thinkers. Mademoiselle Blesseau, his house manager and confidante, summed up Buffon's character in a letter written to his collaborator Faujas de Saint-Fond just after the master's death: "No one has ever been able to take credit for having controlled him." Jacques Roger comments:

> In the hierarchical society in which he lived, he knew how to carve out a place for himself, without excessive qualms or dishonoring servility. He used institutions as he found them and did not seek to change them because it was none of his business and because he did not have a great deal of confidence in human wisdom.

Buffon was just too engaged and too enmeshed to transform the world of thought with a consistent vision—too occupied with his seigneurial rights and funds (where he was fair but demanding, litigious if thwarted, and not particularly kind), and with wheeling and dealing to add land to his estates or to his (and the people's) Parisian gardens and museum. Too busy tending to his household after the early death of his wife, and worrying about his only and wastrel son, who suffered under his father's glory, bearing the diminutive nickname of Buffonet. (After his father's death, Buffonet ended up under the guillotine during the Reign of Terror.) Too involved also in pursuing his own tender, longstanding, and properly discreet relationship with Madame Necker, wife of the finance minister, who comforted and stood by him during his final illness and death. All this hubbub doesn't leave much time, or enough calm and extended space, for developing and propagating a consistent and radical reconstruction of nature.

Buffon's attitude toward religion and his relationships with France's Catholic hierarchy, best illustrate this defining (and ultimately constraining) feature of his personality. He was, without much doubt, a materialist at heart, and at least an agnostic in personal belief. A candid and private remark to Hérault de Séchelles epitomizes both his public stance and his personal attitude: "I have always named the Creator; but we need only remove this word and, of course, put in its place the power of Nature."

Buffon's publications play an extended cat and mouse game with religion. *Histoire naturelle* abounds with flowery and conventional hymns of praise to the omnipotent deity, creator of all things in heaven and earth. But Buffon's content often challenged traditional views and biblical texts. In fact, he began *Natural History* by forthrightly arguing, in volume 1 on the "Theory of the Earth" (published in 1749), that our planet had experienced an unlimited and cyclical history of gradual erosion and exposure of continents, uninterrupted by any catastrophe. (Buffon did not explicitly deny the Noachian deluge, but no one could have missed the implication.)

On January 15, 1751, the Theological Faculty of the Sorbonne attacked Buffon in a strong letter, demanding retraction or censorship. Buffon, in his usual worldly way, backed down in a note of apparent apology, stating that he believed "very firmly all that is told [in the book of Genesis] about Creation, both as to the order of time and the circumstances of the facts," and that he had presented his theory "only as a pure philosophical supposition." Buffon then published the Sorbonne's letter and his response at the beginning of the fourth volume of *Histoire naturelle* in 1753, and in all subsequent editions.

When I was younger, and beguiled by the false myth of warfare between science and religion as the path of progress in Western history, I viewed Buffon's retraction as a sad episode of martyrdom at an intermediary stage along an inevitable road. I now hold an entirely different view. Buffon surely prevailed in this incident. He reached a formal agreement with his enemies, staved off any future attacks, published a meaningless "apology" that no one would regard as sincere, and then never changed a single word of his original text. "It is better to be humble than hung," he wrote to a colleague in describing this contretemps; *Paris vaut bien une messe.*

Nonetheless, as Buffon lay dying, he clamored for last rites with a final ounce of passion that seemed ultimately and poignantly sincere. He had previously said to Hérault de Séchelles, in his usual and somewhat cynical manner, "When I become dangerously ill and feel my end approaching, I will not hesitate to send for the sacraments. One owes it to the public cult." Yet now, faced with the actuality, he seemed to plead only for himself. Madame Necker described his last moments (as a witness, not from secondhand reports): "He spoke to Father Ignace and said to him in a very anxious manner, 'Someone give me the good God quickly! Quickly! Quickly!' . . . Father Ignace gave him communion and M. de Buffon repeated during the ceremony, 'give it then! But give it then!'"

I do not know how to resolve this tangle of complexity, this mixture of practical posturing and sincere conviction. Perhaps we cannot go beyond Jacques Roger's insightful conclusions:

> That Buffon had a passion for order in everything—in his schedule, his accounts, his papers, and his life no less than in his study of nature—was such an obvious aspect of his temperament that his contemporaries noted it. He wanted an order, but not just any order; he wanted a true and legitimate order. Buffon wished there to be an order in society, and . . . he did define a few rules that should preside over such an order. Respect for the established religion is one of them, and he observed it all his life.

If we regard all these foregoing reasons for the eclipse of Buffon's name as primarily "negative" (his failure to construct and defend a transforming and distinctive view of life), another set of factors must be identified as the "positive" fate of all great reformers with such a broad palette and such an immediate impact. First of all, worldliness includes another side that promotes later invisibility. People who build institutions ("brick and mortar" folks, not mere dreamers and schemers), who lobby for educational reform, who write the textbooks that instruct generations of students, become widely known in their generation, if only because they demand explicit obeisance from anyone who wants to engage in the same business. But when they die and no longer hold the strings of power, their names fade quickly from view, even while their institutions and writings continue to mold the history of thought in profound and extensive ways.

Thus, we may note the irony of worldliness in the context of scales of time: one trades immediate recognition in life for the curious status of continuing, but anonymous, influence. How would French biology have developed without the Muséum, and without the forty-four volumes of *Histoire naturelle*? Can a great discovery by a recluse match the ultimately silent achievements of such worldliness? T. H. Huxley, with his tireless round of speeches, exhortations, popular books, politicking, and service on government committees, may have left a greater impact than Darwin upon British society. But Darwin, who, in the last decades of his life, almost never left his country house, even for trips to nearby London, persists (properly, I would claim, in another argument for another time) as the icon of our discoveries and our fears—while Huxley has become a faded memory.

How, similarly, can we measure Buffon's continuing presence? In the recent and brilliant reconstruction of the Grande Galérie of his Muséum into the world's finest modern exhibit on evolution? In *Histoire naturelle,* a treatise that has never been entirely out of print, and that taught students throughout the world as a primary text for more than a century—often in pirated editions that didn't acknowledge Buffon? (For example, few people know, I suspect, that the poet Oliver Goldsmith, to earn his bread, wrote a multivolumed *History of the Earth and Animated Nature* that amounts to little more than lightly annotated Buffon. My own collection of popular science books includes a volume, published in New York during the late nineteenth century, entitled *Buffon's Natural History*—a one-volume amalgam of bits and pieces from *Natural History,* and undoubtedly paying not a penny in royalties to Buffon's estate.)

Yet, and finally, the positive reasons for the paradoxical correlation of later anonymity with continuing impact also include a factor that should be judged as paramount, and that also distills the core of Buffon's greatest contribution to the history of ideas. Some of the grandest tools in the arsenal of our consciousness work so broadly and so generally that we can scarcely assign authorship to a single person. (Darwin can be identified as the discoverer of natural selection, even as the first comprehensive defender of biological evolution based on hard data drawn from all major subjects of natural history. But no one can be called the inventor of a developmental, rather than a static, view of nature.)

Buffon became the central figure in one of the great transformations of human thought—the discovery of history as a guiding principle for organizing the data of the natural world, including many aspects of human diversity (from language, to the arts, to social systems). As the earth's great age—its "deep time"—became apparent, and as revolutionary ideologies replaced monarchies in parts of Europe and America, such a reconstruction of knowledge hovered "in the air," and would have occurred if Buffon had never been born (see Paolo Rossi, *The Dark Abyss of Time,* University of Chicago Press, 1984). But, through a combination of the best subject matter to express such a change, an incomparable prose style, and a wide and dedicated readership, Buffon became the most influential focus of this transformation, with *Histoire naturelle* as the primary agent.

BUFFON'S DISCOVERY AND DEFINITION OF HISTORY

A truly historical account of nature demands deep time. But time can only provide a matrix for the unfolding of events. History requires the ordering of phenomena in narrative form—that is, as a temporal series with direction given by a sequence of complex and unrepeatable events, linked one to the next by sensible reasons for transition. In short, to qualify as history, such a sequence must embody the last two syllables of its name: it must tell a story.

Most pre-Buffonian science included no history. Organisms had been created in primal perfection on a young earth, and none had become extinct (except in the singularity of Noah's flood—and unique transforming events don't constitute a history). The rocks of the earth represented either an original creation or the residues of Noah's flood. Even the influential cosmologies of Newton, and of Buffon's younger and brilliant colleague Laplace, purposely rejected history in positing exactly repeating cycles (perhaps with self-adjusting variations) of "eternal return"—as Darwin recognized so well when he ended his *Origin of Species* (1859) by contrasting the rich historicity of evolution with the sterility of endless cosmological turning: "Whilst this planet has gone cycling on according to the fixed law of gravity, from so simple a beginning endless forms most beautiful and most wonderful have been, and are being, evolved."

In the most important change of his own views between the inception of *Natural History* in 1749 and the publication of its most important volume in 1778, Buffon became an advocate for historical thinking. His first volume of 1749, as discussed earlier, had been sufficiently radical in positing a long and indeterminate age for the earth. Buffon did propose one historical item in this initial *Theory of the Earth*—the first important hypothesis for the origin of planets by cometary impact into the sun, with ejection of masses to form the planetary spheres. But following this tempestuous origin, the earth of Buffon's first volume experienced no further history—for geology had recorded only a series of repeating cycles in erosion and exposure of continents.

But Buffon, confuting the cliché that scientists must develop their best ideas in their youth or not at all, reversed his original belief, devised an intrinsically and thoroughly historical theory of the earth, and published his results at the age of seventy-one, in a volume that became by far his most popular, his most influential, and his most controversial: *Des époques de la nature* (The epochs of nature), which originally appeared in 1778 as supplementary volume five of *Histoire naturelle.* This treatise became the most important scientific document ever written in promoting the transition to a fully historical view of nature. (Since Buffon's influence lay largely in his command of language, *The Epochs of Nature* also illustrates the underappreciated principle that literary style may not be irrelevant to the success of scientific ideas.) And yet, as argued above, this shift to historical thinking raised too big an issue, involving too many subjects and approaches, to lay in one man's lap—so Buffon's name never became firmly attached to his most important intellectual achievement.

The Epochs of Nature rose from complex roots in Buffon's psyche and activities. He did not simply devise this major transition from his armchair. Ever since developing his theory of planetary formation by cometary impact upon the sun, Buffon had searched for evidence that might indicate the time of origin, and the consequent age of the earth. ("Indeterminably long" could not satisfy a man of his restless energy.) After setting up his forges for smelting iron, a testable idea struck Buffon. If the earth had originated as a fireball, he could presumably calculate the length of time required for sufficient cooling to form a solid surface that could serve as a substrate for geological strata and life itself.

Buffon therefore began to experiment with the cooling of iron balls made in his forge. He then scaled his results upward to theoretical calculations for iron balls the size of the earth, and then to more realistic models for balls of various compositions more closely mimicking the earth's structure. Buffon pursued these experiments and calculations for years, obviously enjoying this return to his mathematical roots. He filled chapters of *Histoire naturelle* with his results, and finally decided that the earth must be at least 75,000 years old (and probably a good deal more ancient).

These experiments may have validated deep time in a quantitative manner, but they inspired an even more important change in Buffon's thinking, for they gave him the key to history. A continually cooling earth provided an arrow for time, a fundamental direction for the physical surface and for life as well. Since all organisms originate in perfect adaptation to surrounding environments—and since these environments have changed directionally through time to colder and colder conditions—the composition of faunas must also vary, as some species become extinct when climates alter beyond their power to cope, and new species, adapted to the changed circumstances, then appear.

(As one example of the radical nature of Buffon's historical view, the idea of extinction stuck in the craw of traditional naturalists, who remained committed to an earth made perfect in all ways at the outset, and who therefore could not abide the idea that species might disappear through failure of adaptation. Thomas Jefferson, Buffon's rival, could cite many good reasons for sending Lewis and Clark on their famous expedition, but one small factor lay in his hope that these explorers might find living mammoths in uncharted western lands, thus shaking Buffon's claim that species could die.)

Buffon constructed a rich history of seven successive epochs, all controlled by the continuous cooling of the earth from an original status as a solar fireball: first, the origin of the earth and planets by cometary impact; second, the formation of the solid earth and its mineral deposits; third, the covering of continents by oceans and the origin of marine life; fourth, the retreat of waters and the emergence of new continents; fifth, the appearance of animal life on land; sixth, the fragmentation of continents and the formation of current topography; seventh, the origin of humans and our accession to power.

Note that Buffon did not follow the most traditional arrow of history by arguing that life became continually more complex. In fact, he viewed the first marine creatures of epoch three (including ammonites and fishes) as already fully intricate. Buffon was not, after all, an evolutionist, and he built his arrow of time as a vector of decreasing warmth, not as a rising parade of organic progress. This arrow led him to a pessimistic conclusion, well constructed to fuel cosmic angst: the earth must eventually freeze over, leading to the extirpation of all life. This concept of a "heat death" for the earth became one of the most contentious and interesting ideas in late-eighteenth- and early-nineteenth-century thought, a theme of many poems, plays, and paintings.

Buffon's history also included a set of intriguing consequences, some internal to the theory, others inspired by the reactions of readers. To mention just two, Buffon might be cited by current ecoactivists (and I do say this facetiously) as an antihero—for he developed the notion of a greenhouse effect caused by human burning of forests, but welcomed such an imprint of advancing civilization as a device for postponing the earth's death by cold. Buffon wrote: "Cleansing, clearing, and populating a country gives it heat for several thousand years. . . . Paris and Quebec are at about the same latitude; Paris would therefore be as cold as Quebec if France and all the regions surrounding it were as lacking in men, and covered with woods . . . as are the neighboring lands of Canada."

Second, Buffon became the surprised recipient of several sumptuous gifts from Catherine II of Russia (a.k.a. Catherine the Great)—a collection of furs, all the medals of her reign (in gold), and her portrait on a gold snuffbox encrusted with diamonds. Catherine had been delighted by Buffon's argument that since the earth becomes increasingly colder through time, new species originate in high latitudes and then migrate toward the tropics as temperatures continue to drop. Russia therefore became a cradle of life, rather than the frigid refugium envisioned by most previous writers. Buffon, ingratiating as always, thanked Catherine in a glowing letter that wished her well in campaigns against the Ottoman Empire ("that stagnating part of Europe"), and stated his hope to see "beautiful nature and the arts descend a second time from the North to the South under the standard of [her majesty's] powerful genius."

Moreover, and finally, the eminently orderly Buffon knew exactly what he had accomplished. He consciously promoted history as a novel and coordinating theme for all nature. He not only proposed a theory of origin, an arrow of time, and a narrative in seven epochs. He also knew that the triumph of history would require a fundamentally new way of thinking, and an explicit methodology, not yet familiar to scientists, for reconstructing the immensely long and poorly preserved record of the earth and life. He therefore suggested that natural scientists take their cue from procedures already worked out by students of human history. *The Epochs of Nature* begins with this call for an entirely new mode of thinking:

> In civil history, we consult titles, we research medals, we decipher ancient inscriptions in order to determine the time of human revolutions and to fix the dates of events in the moral order. Similarly, in natural history, it is necessary to excavate the archives of the world, to draw old monuments from the entrails of the earth, to collect their debris, and to reassemble into a single body of proof all the indices of physical changes which enable us to go back to the different ages of nature. This is the only way to fix points in the immensity of space, and to place a certain number of milestones on the eternal route of time. (My translation.)

No other person could possibly have provided better fuel for such a transformation in the history of human thought: this man of such restless energy; this man who operated forges and who developed the experimental and mathematical skill to infer the age of the earth from balls of iron; who composed thirty-six volumes of the greatest treatise ever written in natural history by working fourteen hours a day for more than forty years. And if all these skills and attributes could not turn the tide, Buffon also wrote in an elegant prose that placed him, a "mere" student of nature, among the leading men of letters in his interesting time. Buffon surely knew how to prevail—for style, after all, is the man himself.

5

The Proof of Lavoisier's Plates

I. Writing in the Margins

I ONCE HAD A TEACHER WITH AN IDIOSYNCRATIC HABIT that distressed me forty years ago, but now—and finally, oh sweet revenge!—can work for me to symbolize the general process of human creativity. I never knew a stingier woman, and though she taught history in a junior high school in New York City, she might well have been the frugal New England farmer with a box marked "pieces of string not worth saving." Readers who attended New York public schools in the early 1950s will remember those small yellow slips of paper, three by six inches at most, that served all purposes from spot quizzes to "canvasses" for art class. Well, Mrs. Z. would give us one sheet—only one—for any classroom exam, no matter how elaborate the required answers. She would always reply to any plea for advice about containment or, God forbid, for an additional yellow sheet (comparable in her system of values to Oliver Twist's request for more soup) with a firm refusal followed by a cheery instruction expressed in her oddly lilting voice: ". . . and if you run out of room, just write in the margins!"

Margins play an interesting role in the history of scholarship, primarily for their schizophrenic housing of the two most contradictory forms of intellectual activity. Secondary commentaries upon printed texts (often followed by several layers of commentaries upon the commentaries) received their official designation as "marginalia" to note their necessary position at the edges. The usual status of such discourse as derivative and trivial, stating more and more about less and less at each iteration, leads to the dictionary definition of marginalia as "nonessential items" *(Webster's Third New International)* and inevitably recalls the famous, and literally biting, satire of Jonathan Swift:

> *So, naturalists observe, a flea*
> *Hath smaller fleas that on him prey;*
> *And these have smaller still to bite 'em;*
> *And so proceed* ad infinitum
> *Thus every poet, in his kind,*
> *Is bit by him that comes behind.*

But margins also serve the diametrically opposite purpose of receiving the first fruits and inklings of novel insights and radical revisions. When received wisdom has hogged all the central locations, where else can creative change begin? The curmudgeon and cynic in me regards Thoreau's *Walden* as the most overquoted (and underwhelming) American classic, but I happily succumb for the first time to cite his one-liner for a vibrant existence: "I love a broad margin to my life."

Literal margins, however, must usually be narrow—and some of the greatest insights in the history of human thought necessarily began in such ferociously cramped quarters. The famous story of Fermat's Last Theorem, no matter how familiar, cannot be resisted in this context: when the great mathematician died in 1665, his executors found the following comment in his copy of Diophantus' *Arithmetica,* next to a discussion of the claim that no natural numbers x, y, and z exist such that $x^n + y^n = z^n$, where n is a natural number greater than 2: "I have discovered a truly remarkable proof but this margin is too small to contain it." Mathematicians finally proved Fermat's Last Theorem just a few years ago, to great subsequent fanfare and an outpouring of popular books. But we shall never know if Fermat truly beat the best of the latest by three hundred years, or if (as my own betting money says, admittedly with no good evidence) he had a promising idea and never detected the disabling flaw in the midst of his excitement.

I devote this essay to the happier and opposite story of a great insight that a cramped margin did manage (just barely) to contain and nurture. This tale, for reasons that I do not fully understand, remains virtually unknown (and marginal in this frustrating sense) both to scientists and historians alike—although the protagonist ranks as one of the half dozen greatest scientists in Western history, and the subject stood at the forefront of innovation in his time. In any case, the movement of this insight from marginality in 1760 to centrality by 1810 marks the birth of modern geology, and gives us a rare and precious opportunity to eavesdrop on a preeminent thinker operating in the most exciting and instructive of all times: at a labile beginning in the codification of a major piece of natural knowledge—a unique moment featuring a landscape crossed by one hundred roads, each running in the right general direction toward a genuine truth. Each road, however, reaches a slightly different Rome, and our eventual reading of nature depends crucially upon the initial accidents and contingencies specifying the path actually taken.

In 1700, all major Western scholars believed that the earth had been created just a few thousand years ago. By 1800, nearly all scientists accepted a great antiquity of unknown duration, and a sequential history expressed in strata of the earth's crust. These strata, roughly speaking, form a vertical pile, with the oldest layers on the bottom and the youngest on top. By mapping the exposure of these layers on the earth's surface, this sequential history can be inferred. By 1820, detailed geological maps had been published for parts of England and France, and general patterns had been established for the entirety of both nations. This discovery of "deep time," and the subsequent resolution of historical sequences by geological mapping, must be ranked among the sweetest triumphs of human understanding.

Few readers will recognize the name of Jean-Etienne Guettard (1715–1786), a leading botanist and geologist of his time, and the initiator of the first "official" attempt to produce geological maps of an entire nation. In 1746, Guettard presented a preliminary "mineralogical map" of France to the Académie Royale des Sciences. In subsequent years, he published similar maps of other regions, including parts of North America. As a result, in 1766, the secretary of state in charge of mining commissioned Guettard to conduct a geological survey and to publish maps for all of France. The projected atlas would have included 230 maps, but everyone understood, I suspect, that such a task must be compared with the building of a medieval cathedral, and that no single career or lifetime could complete the job. In 1770, Guettard published the first sixteen maps. The project then became engulfed by political intrigue and finally by a revolution that (to say the least) tended to focus attention elsewhere. Only 45 of the 230 projected maps ever saw the published light of day, and control of the survey had passed to Guettard's opponents by this time.

Guettard's productions do not qualify as geological maps in the modern sense, for he made no effort to depict strata, or to interpret them as layers deposited in a temporal sequence—the revolutionary concepts that validated deep time and established the order of history. Rather, as his major cartographic device, Guettard established symbols for distinctive mineral deposits, rock types, and fossils—and then merely placed these symbols at appropriate locations on his map. We cannot even be sure that Guettard understood the principle of superposition—the key concept that time lies revealed in a vertical layering of strata, with younger layers above (superposed upon) older beds. Guettard did develop a concept of *"bandes,"* or roughly concentric zones of similar rocks, and he probably understood that a vertical sequence of strata might be expressed as such horizontal zones on a standard geographic map. But in any case, he purposely omitted these *bandes* on his maps, arguing that he wished only to depict facts and to avoid theories.

This focus on each factual tree, combined with his studious avoidance of any theoretical forest of generality or explanation, marked Guettard's limited philosophy of science, and also (however unfairly) restricted his future reputation, for no one could associate his name with any advance in general understanding. Rhoda Rappoport, a distinguished historian of science from Vassar College and the world's expert on late-eighteenth-century French geology, writes of Guettard (within a context of general admiration, not denigration): "The talent he most conspicuously lacked was that of generalization, or seeing the implications of his own observations. . . . Most of his work reveals . . . that he tried hard to avoid thinking of the earth as having a history."

But if Guettard lacked this kind of intellectual flair, he certainly showed optimal judgment in choosing a younger partner and collaborator for his geological mapping, for Guettard fully shared this great enterprise with Antoine-Laurent Lavoisier (1743–94), a mere fledgling of promise at the outset of their work in 1766, and the greatest chemist in human history when the guillotine literally cut his career short in 1794.

Guettard and Lavoisier took several field trips together, including a four-month journey in 1767 through eastern France and part of Switzerland. After completing their first sixteen maps in 1770, Lavoisier's interest shifted away from geology toward the sources of his enduring fame—a change made all the more irrevocable in 1777, when control of the geological survey passed to Antoine Monnet, inspector general of mines, and Lavoisier's enemy. (Later editions of the maps ignore Lavoisier's contributions and often don't even mention his name.)

Nonetheless, Lavoisier's geological interests persisted, buttressed from time to time by transient hope that he might regain control of the survey. In 1789, with his nation on the verge of revolution, Lavoisier presented his only major geological paper—a stunning and remarkable work that inspired this essay. Amidst his new duties as *régisseur des poudres* (director of gunpowder), and leading light of the commission that invented the meter as a new standard of measurement—and despite the increasing troubles that would lead to his arrest and execution (for his former role as a farmer-general, or commissioned tax collector)—Lavoisier continued to express his intention to pursue further geological studies and to publish his old results. But the most irrevocable of all changes fractured these plans on May 8, 1794, less than three months before the fall of Robespierre and the end of the Terror. The great mathematician Joseph-Louis Lagrange lamented the tragic fate of his dear friend by invoking the primary geological theme of contrasting time scales: "It took them only an instant to cut off his head, but France may not produce another like it in a century."

All the usual contrasts apply to the team of Guettard and Lavoisier: established conservative and radical beginner; mature professional and youthful enthusiast; meticulous tabulator and brilliant theorist; a counter of trees and an architect of forests. Lavoisier realized that geological maps could depict far more than the mere location of ores and quarries. He sensed the ferment accompanying the birth of a new science, and he understood that the earth had experienced a long history potentially revealed in the rocks of his maps. In 1749, Georges Buffon, the greatest of French naturalists, had begun his monumental treatise (*Histoire naturelle,* which would eventually run to forty-four volumes) with a long discourse on the history and theory of the earth (see chapter 4).

As he groped for a way to understand this history from the evidence of his field trips, and as he struggled to join the insights published by others with his own original observations, Lavoisier recognized that the principle of superposition could yield the required key: the vertical sequence of layered strata must record both time and the order of history. But vertical sequences differed in all conceivable features from place to place—in thickness, in rock types, in order of the layers. How could one take this confusing welter and infer a coherent history for a large region? Lavoisier appreciated the wisdom of his older colleague enough to know that he must first find a way to record and compile the facts of this variation before he could hope to present any general theory to organize his data.

A geological map by Guettard and Lavoisier, with Lavoisier's temporal sequence of strata in the right margin.

Lavoisier therefore suggested that a drawing of the vertical sequence of sediments be included alongside the conventional maps festooned with Guettard's symbols. But where could the vertical sections be placed? In the margins, of course; for no other space existed in the completed design. Each sheet of Guettard and Lavoisier's *Atlas* therefore features a large map in the center with two marginal columns on the side: a tabular key for Guettard's symbols at the left, and Lavoisier's vertical sections on the right. If I wished to epitomize the birth of modern geology in a single phrase (admittedly oversimplified, as all such efforts must be), I would honor the passage—both conceptual and geometric—of Lavoisier's view of history, as revealed in sequences of strata, from a crowded margin to the central stage.

Many fundamental items in our shared conceptual world seem obvious and incontrovertible only because we learned them (so to speak) in our cradle and have never even considered that alternatives might exist. We often regard such notions—including the antiquity of the earth, the rise of mountains, and the deposition of sediments—as simple facts of observation, so plain to anyone with eyes to see that any other reading could only arise from the province of knaves or fools. But many of these "obvious" foundations arose as difficult and initially paradoxical conclusions born of long struggles to think and see in new ways.

If we can recapture the excitement of such innovation by temporarily suppressing our legitimate current certainties, and reentering the confusing transitional world of our intellectual forebears, then we can understand why all fundamental scientific innovation must marry new ways of thinking with better styles of seeing. Neither abstract theorizing nor meticulous observation can provoke a change of such magnitude all by itself. And when—as in this story of Lavoisier and the birth of geological mapping—we can link one of the greatest conceptual changes in the history of science with one of the most brilliant men who ever graced the profession, then we can only rejoice in the enlarged insight promised by such a rare conjunction.

Most of us, with minimal training, can easily learn to read the geological history of a region by studying the distribution of rock layers on an ordinary geographic map and then coordinating this information with vertical sections (as drawn in Lavoisier's margins) representing the sequence of strata that would be exposed by digging a deep hole in any one spot. But consider, for a moment, the intellectual stretching thus required, and the difficulty that such an effort would entail if we didn't already understand that mountains rise and erode, and that seas move in and out, over any given region of our ancient earth.

A map is a two-dimensional representation of a surface; a vertical section is a one-dimensional listing along a line drawn perpendicular to this surface and into the earth. To understand the history of a region, we must mentally integrate these two schemes into a three-dimensional understanding of time (expressed as vertical sequences of strata) across space (expressed as horizontal exposures of the same strata on the earth's surface). Such increases in dimensionality rank among the most difficult of intellectual problems—as anyone will grasp by reading the most instructive work of science fiction ever published, E. A. Abbott's *Flatland* (originally published in 1884 and still in print), a "romance" (his description) about the difficulties experienced by creatures who live in a two-dimensional world when a sphere enters the plane of their entire existence and forces them to confront the third dimension.

As for the second component of our linkage, I can only offer a personal testimony. My knowledge of chemistry remains rudimentary at best, and I can therefore claim no deep understanding of Lavoisier's greatest technical achievements. But I have read several of his works and have never failed to experience one of the rarest emotions in my own arsenal: sheer awe accompanied by spinal shivers. A kind of eerie, pellucid clarity pervades Lavoisier's writing (and simply makes me ashamed of the peregrinations in these essays).

Perhaps, indeed almost certainly, a few other scientists have combined equal brilliance with comparable achievement, but no one can touch Lavoisier in shining a light of logic into the most twisted corners of old conceptual prisons, into the most tangled masses of confusing observations—and extracting new truths expressed as linear arguments accessible to anyone. As an example of the experimental method in science (including the fundamental principle of double-blind testing), no one has ever bettered the document that Lavoisier wrote in 1784 as head of a royal commission (including Benjamin Franklin, then resident in Paris and, ironically, Dr. Guillotin, whose "humane" invention would end Lavoisier's life) to investigate (and, as results proved, refute) the claims of Dr. Mesmer about the role of animal magnetism in the cure of disease by entrancement (mesmerization).

Lavoisier did not compose his only geological paper until 1789, but Rhoda Rappoport has shown that he based this work upon conclusions reached during his mapping days with Guettard. Lavoisier did not invent the concept of vertical sections; nor did he originate the idea that sequences of strata record the history of regions on an earth of considerable antiquity. Instead, he resolved an issue that may seem small by comparison, but that couldn't be more fundamental to any hope for a workable science of geology (as opposed to the simpler pleasures of speculating about the history of the earth from an armchair): he showed how the geological history of a region can be read from variation in strata from place to place—or, in other words, how a set of one-dimensional lists of layered strata at single places can be integrated by that greatest of all scientific machines, the human mind, into a three-dimensional understanding of the history of geological changes over an entire region.

(I doubt that Lavoisier's work had much actual influence, for he published only one paper on this subject and did not live to realize his more extensive projects. Other investigators soon reached similar conclusions, for the nascent science of geology became the hottest intellectual property in late-eighteenth-century science. Lavoisier's paper has therefore been forgotten, despite several efforts by isolated historians of science through the years, with this essay as the latest attempt, to document the singularity of Lavoisier's vision and accomplishment.)

From my excellent sample of voluminous correspondence with lay readers during a quarter century of writing these essays, I have grasped the irony of the most fundamental misunderstanding about science among those who love the enterprise. (I am not discussing the different errors made by opponents of science.) Supporters assume that the greatness and importance of a work correlates directly with its stated breadth of achievement: minor papers solve local issues, while great works fathom the general and universal nature of things. But all practicing scientists know in their bones that successful studies require strict limitation: one must specify a particular problem with an accessible solution, and then find a sufficiently simple situation where attainable facts might point to a clear conclusion. Potential greatness then arises from cascading implications toward testable generalities. You don't reach the generality by direct assault without proper tools. One might as well dream about climbing Mount Everest in a T-shirt, wearing tennis shoes, and with a backpack containing only an apple and a bottle of water.

II. Capturing the Center

When Lavoisier began his geological work with Guettard in 1766, he accepted a scenario, then conventional, for the history of the earth as revealed by the record of rocks: a simple directional scheme that envisaged a submergence of ancient landmasses (represented today by the crystalline rocks of mountains) under an ocean, with all later sediments formed in a single era of deposition from this stationary sea (on this topic, see Rhoda Rappoport's important article "Lavoisier's Theory of the Earth," *British Journal for the History of Science,* 1973). Since geologists then lacked techniques for unraveling the contorted masses of older crystalline rocks, they devoted their research to the later stratified deposits, and tried to read history as an uncomplicated tale of linear development. (No fossils had been found in the older crystalline rocks, so early geologists also assumed that the later stratified deposits contained the entire history of life.)

Lavoisier's key insight led him to reject this linear view (one period of deposition from a stationary sea) and to advocate the opposite idea that sea level had oscillated through time, and that oceans had therefore advanced and retreated through several cycles in any particular region—a notion now so commonplace that any geologist can intone the mantra of earth history: "the seas go in and the seas go out." Lavoisier reached this radical conclusion by combining the developing ideas of such writers as Buffon and De Maillet with his own observations on cyclical patterns of sedimentation in vertical sections.

Lavoisier christened his 1789 paper with a generous title characteristic of a time that did not separate literature and science: *Observations générales sur les couches modernes horizontales qui ont été déposées par la mer, et sur les conséquences qu'on peut tirer de leurs dispositions relativement à l'ancienneté du globe terrestre* (General observations on the recent horizontal beds that have been deposited by the sea, and on the consequences that one can infer, from their arrangement, about the antiquity of the earth). Lavoisier's title may be grand, general, and expansive, but his content remained precise, local, and particular—at first! Lavoisier begins his treatise by distinguishing the properties of sediments deposited in open oceans from those formed along shorelines—a procedure that he then followed to build the data for his central argument that seas advance and retreat in a cyclical pattern over any given region.

After two short introductory paragraphs, Lavoisier plunges right in by expressing puzzlement that two such opposite kinds of rock so often alternate to form multiple cycles in a single vertical section. Criteria of fossils and sediment indicate calm and gentle deposition for one kind: "Here one finds masses of shells, mostly thin and fragile, and most showing no sign of wear or abrasion. . . . All the features [of the rocks] that surround these shells indicate a completely tranquil environment" (my translations from Lavoisier's 1789 paper). But rocks deposited just above testify to completely different circumstances of formation: "A few feet above the place where I made these observations, I noted an entirely opposite situation. One now sees no trace of living creatures; instead, one finds rounded pebbles whose angles have been abraded by rapid and long-continued tumbling. This is the picture of an agitated sea, breaking against the shore, and violently churning a large quantity of pebbles." Lavoisier then poses his key question, already made rhetorical by his observations:

> How can we reconcile such opposite observations? How can such different effects arise from the same cause? How can movements that have abraded quartz, rock crystal, and the hardest stones into rounded pebbles, also have preserved light and fragile shells?

The simple answer to this specific and limited question may then lead to important generalities for the science of geology, and also to criteria for unraveling the particular history of the earth:

> At first glance, this contrast of tranquillity and movement, of organization and disorder, of separation and mixture, seemed inexplicable to me; nevertheless, after seeing the same phenomena again and again, at different times and in different places, and by combining these facts and observations, it seemed to me that one could explain these striking observations in a simple and natural manner that could then reveal the principal laws followed by nature in the generation of horizontal strata.

Lavoisier then presents his idealized model of a two-stage cycle as an evident solution to this conundrum:

> Two kinds of very distinct beds must exist in the mineral kingdom: one kind formed in the open sea . . . which I shall call *pelagic* beds, and the other formed at the coast, which I shall call *littoral* beds.

Pelagic beds arise by construction, as "shells and other marine bodies accumulate slowly and peacefully during an immense span of years and centuries." But littoral beds, by contrast, arise by "destruction and tumult . . . as parasitic deposits formed at the expense of coastlines."

In a brilliant ploy of rhetoric and argument, Lavoisier then builds his entire treatise as a set of consequences from this simple model of two types of alternating sediments, representing the cycle of a rising and falling sea. This single key, Lavoisier claims, unlocks the great conceptual problem of moving from one-dimensional observations of vertical sequences in several localities to a three-dimensional reconstruction of history. (I call the solution three-dimensional for a literal reason, emphasized earlier in this essay in my discussion of geological maps: the two horizontal dimensions record geographic variation over the earth's surface, while the vertical dimension marks time in a sequence of strata):

> This distinction between two kinds of beds . . . suddenly dispersed the chaos that I experienced when I first observed terranes made of horizontal beds. This same distinction then led me to a series of consequences that I shall try to convey, in sequence, to the reader.

The remainder of Lavoisier's treatise presents a brilliant fusion of general methodology and specific conclusions, a combination that makes the work such a wonderful exemplar of scientific procedure at its best. The methodological passages emphasize two themes: the nature of proof in natural history, and the proper interaction of theory and observation. Lavoisier roots the first theme in a paradox presented on pages 98–99: the need to simplify at first in order to generalize later. Science demands repetition for proper testing of observations—for how else could we learn that the same circumstances reliably generate the same results? But the conventional geologies of Lavoisier's time stymied such a goal—for one directional period of deposition from a single stationary sea offered no opportunity for testing by repetition. By contrast, Lavoisier's model of alternating pelagic and littoral beds provided a natural experiment in replication at each cycle.

But complex nature defies the needs of laboratory science for simple and well-controlled situations, where events can be replicated under identical conditions set by few variables. Lavoisier argues that we must therefore try to impose similar constraints upon the outside world by seeking "natural experiments" where simple models of our own construction might work adequately in natural conditions chosen for their unusual clarity and minimal number of controlling factors.

Consider three different principles, each exploited by Lavoisier in this paper, for finding or imposing a requisite simplicity upon nature's truly mind-boggling complexity.

1. Devise a straightforward and testable model. Lavoisier constructed the simplest possible scheme of seas moving in and out, and depositing only two basic (and strongly contrasting) types of sediment. He knew perfectly well that real strata do not arrange themselves in neat piles of exactly repeating pairs, and he emphasized two major reasons for nature's much greater actual complexity: first, seas don't rise and fall smoothly, but rather wiggle and jiggle in small oscillations superposed upon any general trend; second, the nature of any particular littoral deposit depends crucially upon the type of rock being eroded at any given coastline. But Lavoisier knew that he must first validate the possibility of a general enterprise—three-dimensional reconstruction of geological history—by devising a model that could be tested by replication. The pleasure of revealing unique details would have to wait for another time. He wrote:

> Beds formed along the coast by a rising sea will have unique characteristics in every different circumstance. Only by examining each case separately, and by discussing and explaining them in comparison to each other, will it be possible to grasp the full range of phenomena. . . . I will therefore treat [these details] in a separate memoir.

2. Choose a simple and informative circumstance. Nature's inherent complexity of irreducible uniqueness for each object must be kept within workable scientific bounds by intelligent choice of data with unusual and repeated simplicity. Here Lavoisier lucked out. He had noted the problem of confusing variation in littoral deposits based on erosion of differing rocks at varying coasts. Fortunately, in the areas he studied near Paris, the ancient cliffs that served as sources for littoral sediments might almost have been "made to order" for such a study. The cliffs had been formed in a widespread deposit of Cretaceous age called *La Craie* (or "the Chalk"—the same strata that build England's White

Cliffs of Dover). The Chalk consists primarily of fine white particles, swiftly washed out to sea as the cliffs erode. But the Chalk also includes interspersed beds of hard flint nodules, varying in size from golf balls to baseballs in most cases. These nodules provide an almost perfect experimental material (in uniform composition and limited range of size) for testing the effects of shoreline erosion. Lavoisier noted in particular (I shall show his engravings later in this essay) that the size and rounding of nodules should indicate distance of deposition from the shoreline—for pebbles should be large and angular when buried at the coast (before suffering much wear and erosion), but should then become smaller and rounder as they tumble farther away from the coastline in extensive erosion before deposition.

3. Ask a simple and resolvable question. You needn't (and can't) discover the deep nature of all reality in every (or any!) particular study. Better to pose smaller, but clearly answerable, questions with implications that then cascade outward toward a larger goal. Lavoisier had devised a simple, and potentially highly fruitful, model of oscillating sea levels in order to solve a fundamental question about the inference of a region's geological history from variation in vertical sections from place to place—the sections that he had placed in the margins of the maps he made with Guettard. But such a model could scarcely fail to raise, particularly for a man of Lavoisier's curiosity and brilliance, the more fundamental question—a key, perhaps, to even larger issues in physics and astronomy—of why oceans should rise and fall in repeated cycles. Lavoisier noted the challenge and wisely declined, recognizing that he was busy frying some tasty and sizable fish already, and couldn't, just at the moment, abandon such a bounty in pursuit of Moby Dick. So he praised his work in progress and then politely left the astronomical question to others (although he couldn't resist the temptation to drop a little hint that might help his colleagues in their forthcoming labors):

> It would be difficult, after such perfect agreement between theory and observation—an agreement supported at each step by proofs obtained from strata deposited by the sea—to claim that the rise and fall of the sea [through time] is only a hypothesis and not an established fact derived as a direct consequence of observation. It is up to the geometers, who have shown such wisdom and genius in different areas of physical astronomy, to enlighten us about the cause

> of these oscillations [of the sea], and to teach us if they are still occurring, or if it is possible that the earth has now reached a state of equilibrium after such a long sequence of centuries. Even a small change in the position of the earth's axis of rotation, and a consequent shift in the position of the equator, would suffice to explain all these phenomena. But this great question belongs to the domain of physical astronomy, and is not my concern.

For the second methodological theme of interaction between observation and theory in science, Lavoisier remembered the negative lesson that he had learned from the failures of his mentor Guettard. A major, and harmful, myth of science—engendered by a false interpretation of the eminently worthy principle of objectivity—holds that a researcher should just gather facts in the first phase of study, and rigorously decline to speculate or theorize. Proper explanations will eventually emerge from the data in any case. In this way, the myth proclaims, we can avoid the pitfalls of succumbing to hope or expectation, and departing from the path of rigorous objectivity by "seeing" only what our cherished theory proclaims as righteous.

I do appreciate the sentiments behind such a recommendation, but the ideal of neutrally pure observation must be judged as not only impossible to accomplish, but actually harmful to science in at least two major ways. First, no one can make observations without questions in mind and suspicions about forthcoming results. Nature presents an infinity of potential observations; how can you possibly know what might be useful or important unless you are seeking an answer to a particular puzzle? You will surely waste a frightful amount of time when you don't have the foggiest idea about the potential outcomes of your search.

Second, the mind's curiosity cannot be suppressed. (Why would anyone ever want to approach a problem without this best and most distinctive tool of human uniqueness?) Therefore, you will have suspicions and preferences whether you acknowledge them or not. If you truly believe that you are making utterly objective observations, then you will easily tumble into trouble, for you will probably not recognize your own inevitable prejudices. But if you acknowledge a context by posing explicit questions to test (and, yes, by inevitably rooting for a favored outcome), then you will be able to specify—and diligently seek, however much you may hope to fail—the observations that can refute your preferences. Objectivity cannot be equated with mental blankness; rather, objectivity resides in recognizing your preferences and then subjecting them to especially harsh scrutiny—and also in a willingness to revise or abandon your theories when the tests fail (as they usually do).

Lavoisier had spent years watching Guettard fritter away time by an inchoate gathering of disparate bits of information, without any cohesive theory to guide and coordinate his efforts. As a result, Lavoisier pledged to proceed in an opposite manner, while acknowledging that the myth of objectivity had made his procedure both suspect and unpopular. Nonetheless, he would devise a simple and definite model, and then gather field observations in a focused effort to test his scheme. (Of course, theory and observation interact in subtle and mutually supporting ways. Lavoisier used his preliminary observations to build his model, and then went back to the field for extensive and systematic testing.) In an incisive contrast between naive empiricism and hypothesis testing as modes of science, Lavoisier epitomized his preference for the second method:

> There are two ways to present the objects and subject matter of science. The first consists in making observations and tracing them to the causes that have produced them. The second consists in hypothesizing a cause, and then seeing if the observed phenomena can validate the hypothesis. This second method is rarely used in the search for new truths, but it is often useful for teaching, for it spares students from difficulties and boredom. It is also the method that I have chosen to adopt for the sequence of geological memoirs that I shall present to the Academy of Sciences.

Lavoisier therefore approached the terranes of France with a definite model to test: seas move in and out over geographic regions in cycles of advancing and retreating waters. These oscillations produce two kinds of strata: pelagic deposits in deeper waters and littoral deposits fashioned from eroded coasts near the shoreline. Type of sediment should indicate both environment of deposition and geographic position with respect to the shoreline at any given time: pelagic deposits always imply a faraway shore. For the nearer littoral deposits, relative distance from shore can be inferred from the nature of any particular stratum. For littoral beds made mainly of flint nodules derived from Chalk, the bigger and more angular the nodules, the closer the shoreline.

From these simple patterns, all derived as consequences of an oscillating sea, we should be able to reconstruct the three-dimensional history of an entire region from variation in vertical sequences of sediments from place to place. (For example, if a continuous bed representing the same age contains large and angular flint nodules at point A, and smaller and more rounded nodules at point B, then A lay closer to the shoreline at the time of deposition.)

Lavoisier devotes most of his paper, including all of his seven beautifully drafted plates, to testing this model, but I can summarize this centerpiece of his treatise in three pictures and a few pages of text because the model makes such clear and definite predictions—and nature must either affirm or deny. Lavoisier's first six plates—in many ways, the most strikingly innovative feature of his entire work—show the expected geographic distribution of sediments under his model.

Lavoisier's first plate, for example, shows the predictable geographic variation in a littoral bed formed by a rising sea. The sea will mount from a beginning position (marked *"ligne de niveau de la basse mer,"* or line of low sea level, and indicated by the top of the illustrated waters) to a highstand marked *"ligne de niveau de la haute mer,"* or line of high sea level. The rising sea beats against a cliff, shown at the far left and marked *"falaise de Craye avec cailloux,"* or cliff of Chalk with pebbles. Note that, as discussed previously, this Chalk deposit contains several beds of flint nodules, drawn as thin horizontal bands of dark pebbles.

The rising sea erodes this cliff and deposits a layer of littoral beds underneath the waters, and on top of the eroded chalk. Lavoisier marks this bed with a sequence of letters BDFGHILMN, and shows how the character of the sediment varies systematically with distance from the shoreline. At B, D, and F, near the shore, large and angular pebbles (marked *"cailloux roulés,"* or rolled pebbles), formed from the eroded flint nodules, fill the stratum. The size of particles then decreases continually away from shore, as the pebbles break up and erode (changing from *"sable grossier,"* or coarse sand, to *"sable fin,"* or fine sand, to *"sable très fin ou argille,"* or very fine sand and clay). Meanwhile, far from shore (marked KK at the right of the figure), a pelagic bed begins to form (marked *"commencement des bancs calcaires,"* or beginning of calcareous beds).

From this model, Lavoisier must then predict that a vertical section at G, for example, would first show (as the uppermost stratum) a littoral bed made of large and angular pebbles, while a vertical section at M would feature a pelagic bed on top of a littoral bed, with the littoral bed now made of fine sand and clay. The two littoral beds at G and M would represent the same time, but their differences in composition would mark their varying distances from the shore. This simple principle of relating differences in beds of the same age to varying environments of deposition may seem straightforward, but geologists did not really develop a usable and consistent theory of such "facies" (as we call these variations) until this century. Lavoisier's clear vision of 1789, grossly simplified though his example may be, seems all the more remarkable in this context.

Lavoisier's first plate, showing spatial variation in sediments deposited in a rising sea.

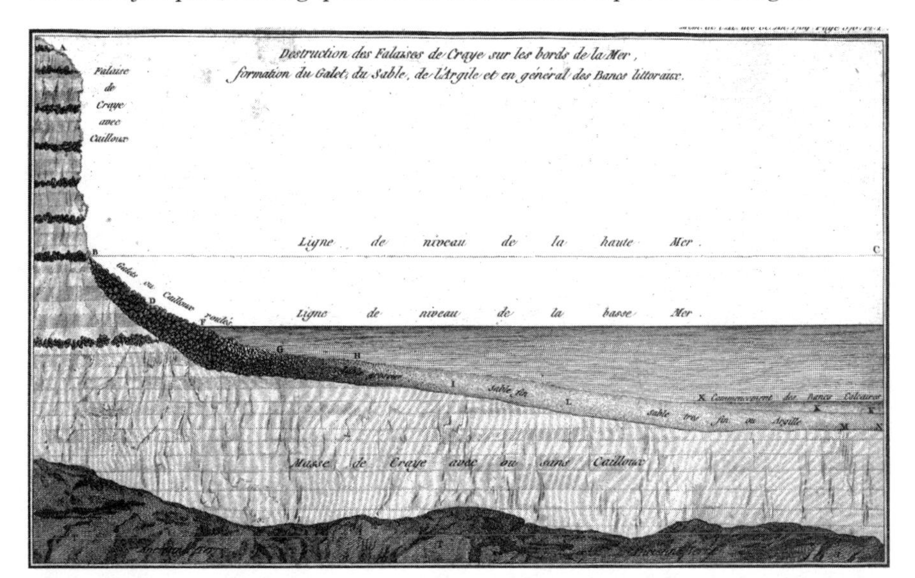

Lavoisier then presents a series of similar diagrams of growing complexity, culminating in plate 6, also reproduced here. This final plate shows the results of a full cycle—the sea, having advanced to its full height, has already retreated back to its starting point. The chalk cliff has been completely eroded away and now remains only as a bottom layer. (Note the distinctive bands of flint nodules for identification. I will discuss later the lowermost layer, marked *"ancienne terre,"* or ancient earth.) A lower littoral layer, formed by the rising sea, lies above the eroded chalk (marked HLMN as *"bancs littoraux inférieurs formés à la mer montante,"* or lower littoral beds formed by the rising sea). A pelagic bed, marked KKK (don't blame Lavoisier for a later and fortuitous American anachronism!), lies just above (labeled *"bancs pelagiens calcaires horizontaux supérieurs,"* or upper calcareous horizontal pelagic beds). Note how the pelagic bed pinches out toward shore because sediments of this type can be deposited only in deep water. This pelagic bed forms when sea level reaches its highest point. Then, as the sea begins to fall, another littoral bed will be deposited in progressively shallower water atop the pelagic bed (marked HIGG and labeled *"bancs littoraux supérieurs formés à la mer déscendante,"* or upper littoral beds formed by the falling sea).

Again, Lavoisier's insights are subtle and detailed—and several specific predictions can be derived from his model. For example, the upper and lower littoral beds will be confluent near the coast because the intervening pelagic bed didn't reach this far inland. Thus, a vertical section taken near the coast should show a single thick littoral bed made of large and angular pebbles. But farther away from shore, a vertical section should include a full array of alternating beds, illustrating the entire cycle and moving (top to bottom, as shown by the vertical line, located just left of center and marked 12345) from the upper littoral bed of the falling sea (1), to the intervening pelagic bed (2), to the lower littoral bed of the rising sea (3), to the underlying chalk (4), and finally to the foundation of the *ancienne terre* (5).

Thus, Lavoisier's model makes clear and specific predictions about how the sediments deposited in full cycles of rising and falling seas should be expressed in the vertical sections that had once adorned the margins of his maps with Guettard, and that represented his signal and original contribution to the developing science of geology. Moreover, the model specified predictions not only for vertical sequences in single places, but also for geographic variation in sequences from place to place. Therefore, in a last figure, Lavoisier presents some actual vertical sections measured in the field. The example presented here corresponds exactly to his prediction for section 12345 in the idealized model. Note the perfect correspondence between Lavoisier's *"Coupe des Montagnes des environ de St. Gobain"* (section through the mountains in the neighborhood of St. Gobain) and his model (except that the actual section doesn't extend below the chalk into the ancient basement). The measured section shows four layers labeled "upper littoral," "pelagic beds," "lower littoral," and "chalk" (note the layers of flint nodules in the lowermost chalk). Lavoisier had intended to write several more geological papers filled with similar empirical details to test his model. Thus, this pilot study presents only a few actual sections, but with impressive promise for continuing validation. Lavoisier had achieved a scientific innovation of the finest and most indubitable form: he had added a dimension (literally) to our knowledge of natural history.

As if he had not done enough already, Lavoisier then ended his treatise with two pages of admittedly hypothetical reasoning on the second great general theme in the study of time and history. His model of oscillating seas lies fully within the Newtonian tradition of complete and ahistorical generality. Lavoisier's oceanic cycles operate through time, but they do not express history because strings of events never occur in distinct and irreversibly directional sequences, and no single result ever denotes a uniquely definable moment. The cycles obey a timeless law of nature, and proceed in the same way no matter when they run. Cycle 100 will yield the same results as cycle 1; and the record

Lavoisier's final plate, showing the spatial and temporal complexity of sediments deposited in a full cycle of a rising and falling sea.

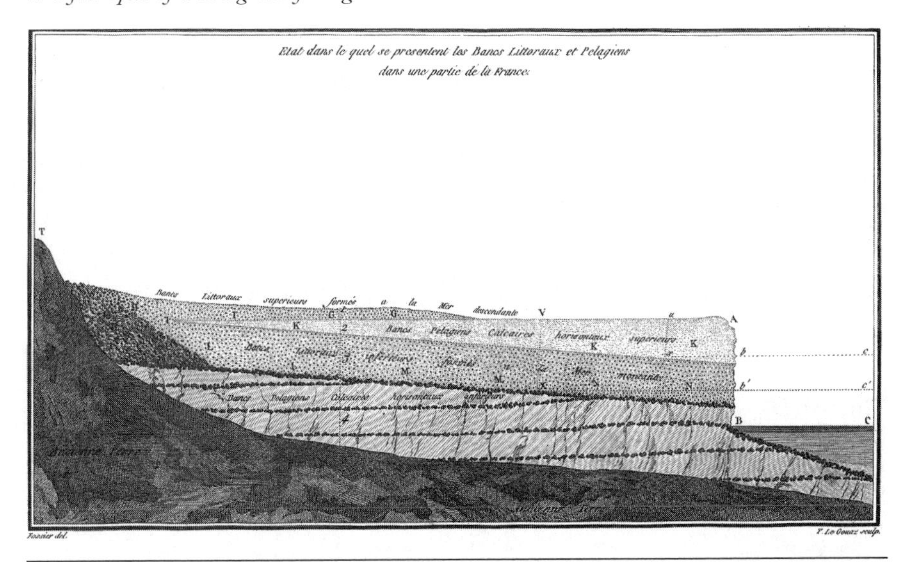

of rocks can never tell us where we stand in the flow of history. All variation reflects either a general environment (high or low sea) or a local circumstance (type of rock in the cliff being eroded), not the distinctive imprint of any unique and definite historical event.

Lavoisier, in other words, had worked brilliantly with the necessary concept of "time's cycle," so vital for any scientific account of the past because we need general laws to explain repeated physical events. But geology cannot render a full account of the earth's past without also invoking the fundamentally different, but intricately conjoined and equally necessary, concept of "time's arrow," so indispensable because geology embraces history as well—and historical accounts must tell stories defined by directional sequences of unique events.

As a prerequisite for interest and meaning, history must require a matrix of extensive time—which Lavoisier had already provided by combining his oscillating model of the oceans with empirical evidence for multiple cycles in vertical sections. If each cycle required considerable time (particularly for the formation of pelagic beds, so slowly built from the debris of organisms), then the evidence for numerous cycles implied an earth of great antiquity. By 1789 (and contrary to popular legend), few scientists still accepted a biblical chronology of just a few thousand years for the earth's history. But the true immensity of geological time still posed conceptual difficulties for many investigators, and Lavoisier's forthright claims mirrored the far more famous lines, published just

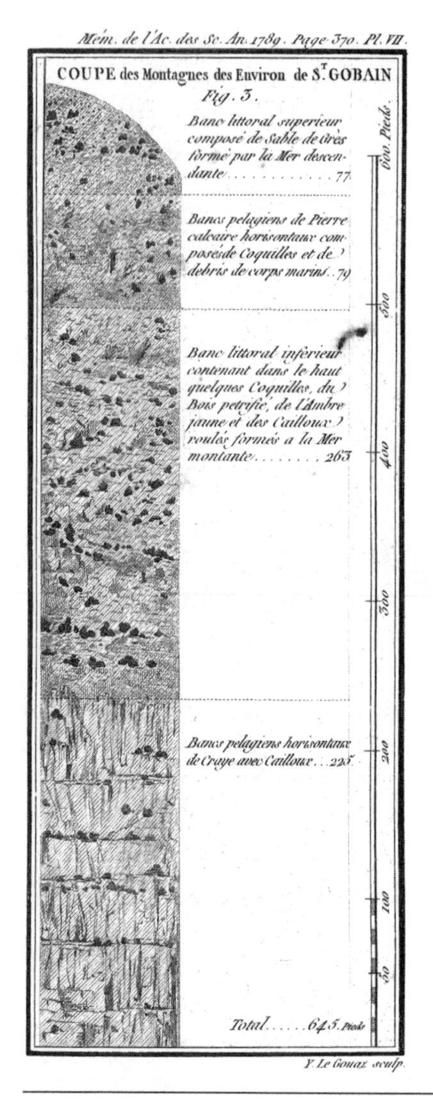

Lavoisier shows that an actual sequence of sediments conforms to his model of deposition in cycles of raising and lowering of sea level.

a year before in 1788, by the traditional "father" of modern geology, the Scotsman James Hutton: "time is, to nature, endless and as nothing." Lavoisier expressed his version of deep time in the more particular light of his model:

> The details that I have just discussed have no other object than to prove this proposition: if we suppose that the sea undergoes a very slow oscillatory movement, a kind of flux and reflux, that these movements occur during a period of several hundreds of thousands

> of years, and that these movements have already occurred a certain number of times, then if we make a vertical section of rocks deposited between the sea and the high mountains, this section must present an alternation of littoral and pelagic beds.

Within such a matrix of deep time, the concept of a truly scientific history obtains new meaning and promise. At the end of his treatise, Lavoisier therefore touches upon this subject in his characteristically empirical way—by returning to the lowermost layer beneath the recorded sediments of his models and measured sections, a complex of rocks that he had bypassed with the simple label *"ancienne terre,"* or ancient earth. Lavoisier now states that he does not regard this foundation as part of the original earth at its time of formation, but rather as a probable series of sediments, much older than the Chalk, but also built as a sequence of littoral and pelagic beds (although now hard to identify because age has obliterated the characteristic features of such deposits):

> One will no doubt want to know about the rocks found underneath the Chalk, and what I mean by the expression *l'ancienne terre.* . . . This is almost surely not the original earth; on the contrary, it appears that what I have called *ancienne terre* is itself composed of littoral beds much older than those depicted in the figures.

In a remarkable passage, Lavoisier then invokes what would become the classic subject for juxtaposing the yin of history (or time's arrow) against the yang of constant features built by invariant laws (time's cycle) to form a complete science of geology: the directional character of life's pageant, the primary component of the earth's rip-roaring narrative story. (By the way, Lavoisier's particular claims turn out to be wrong in every detail, but I can hardly think of an observation more irrelevant to my present point. In 1789, no one knew much about paleontological particulars. I am stressing Lavoisier's keen and correct vision that life would provide the primary source of directional history, or time's arrow.)

Lavoisier bases his claim for history upon a clever argument. He believes that rocks of the *ancienne terre* contain no fossils. But if these rocks include (as he has just argued) the same alternation of pelagic and littoral beds found in younger sediments, then the invariant physical laws of time's cycle should lead us to expect fossils in these strata—for such sediments form in environments that now teem with life. Therefore, time's arrow of directional history must explain the difference. Physical conditions of the *ancienne terre* could not have differed from later circumstances that generated similar sediments, but the earth must then have housed no living creatures if these identical rocks contain no fossils.

Lavoisier then argues that sediments occasionally found below the Chalk (the oldest rocks with marine fossils), but above the *ancienne terre,* often contain fossils of plants. He therefore envisages a threefold directional history of life: an original earth devoid of organisms, followed by the origin of vegetation on land, and finally culminating in the development of animal life both in the sea and on land:

> It is very remarkable that the Chalk is usually the youngest rock to contain shells and the remains of other marine organisms. The beds of shale that we sometimes find below the Chalk often include vestiges of floating bodies, wood, and other vegetable matter thrown up along the coasts. . . . If we may be allowed to hazard a guess about this strange result, I believe we might be able to conclude, as Mr. Monge has proposed [the important French mathematician Gaspard Monge, who served with Lavoisier on the revolutionary commission to devise the metric system], that the earth was not always endowed with living creatures, that it was, for a long time, an inanimate desert in which nothing lived, that the existence of vegetables preceded that of most animals, or at least that the earth was covered by trees and plants before the seas were inhabited by shellfish.

And thus, hurriedly, at the very end of a paper intended only as a preliminary study, an introductory model to be filled in and fleshed out with extensive data based on field research, Lavoisier appended this little conjectural note—to show us, I suspect, that he grasped the full intellectual range of the problems set by geology, and that he recognized the power of combining a firm understanding of timeless and invariant laws with a confident narration of the rich directional history of an ancient earth. His last page bubbles with enthusiasm for future plans involving the whole earth, a project so soon cut off by the evil that only men can do. Consider the poignant paragraph just following his speculation about the history of life:

> In the next article, I will discuss in very great detail these opinions, which really belong more to Mr. Monge than to myself. But it is

> indispensable that I first establish, in a solid way, the observations on which they are based.

I don't know why Lavoisier's execution affects me so deeply. We cannot assert with confidence that he would have completed his geological projects if he had lived (for all creative careers remain chock full of unrealized plans); and we know that he faced his end with a dignity and equanimity that can still provide comfort across the centuries. He wrote in a last letter:

> I have had a fairly long life, above all a very happy one, and I think that I shall be remembered with some regrets and perhaps leave some reputation behind me. What more could I ask? The events in which I am involved will probably save me from the troubles of old age. I shall die in full possession of my faculties.

Lavoisier needs no rescue, either from me or from any modern author. Yet speaking personally (a happy privilege granted to essayists ever since Montaigne invented the genre for this explicit purpose more than two hundred years before Lavoisier's time), I do long for some visceral sense of fellowship with this man who stands next to Darwin in my private pantheon of intellectual heroes. He died through human cruelty, and far too young. His works, of course, will live—and he needs no more. But, and I have no idea why, we also long for what I called visceral fellowship just above—some sense of *physical continuity,* some sign of an *actual presence* to transmit across the generations, so that we will not forget the person behind the glorious ideas. (Perhaps my dedication to such material continuity only marks a personal idiosyncrasy—but not, I think, a rare feeling, and certainly concentrated among those who choose paleontology for a profession because they thrill to the objective records of life's continuous history.)

So let me end with a personal testimony. Through the incredible good fortune of an odd coincidence in good timing and unfathomable pricing, I was able to buy a remarkable item at auction a while ago—the original set of proof plates, each personally signed by Lavoisier, of the seven figures (including the three reproduced here) that accompany his sole geological article of 1789. Two men signed each plate: first, in a thick and bold hand, Gabriel de Bory, vice-secretary of the Academy of Sciences (signed "Bory Vice-Secretaire"); and second, in a much more delicate flow composed of three flourishes surrounding the letters of his last name alone, Antoine-Laurent Lavoisier.

Lavoisier's signature (left) on one of his geological plates.

Lavoisier's own flourishes enhance the visual beauty of the plates that express the intellectual brilliance of his one foray into my field of geology—all signed in the year of the revolution that he greeted with such hope (and such willingness to work for its ideals); the revolution that eventually repaid his dedication in the most cruel of all possible ways. But now I hold a tiny little bit, only a symbol really, of Lavoisier's continuing physical presence in my professional world.

The skein of human continuity must often become this tenuous across the centuries (hanging by a thread, in the old cliché), but the circle remains unbroken if I can touch the ink of Lavoisier's own name, written by his own hand. A candle of light, nurtured by the oxygen of his greatest discovery, never burns out if we cherish the intellectual heritage of such unfractured filiation across the ages. We may also wish to contemplate the genuine physical thread of nucleic acid that ties each of us to the common bacterial ancestor of all living creatures, born on Lavoisier's *ancienne terre* more than 3.5 billion years ago—and never since disrupted, not for one moment, not for one generation. Such a legacy must be worth preserving from all the guillotines of our folly.

6

A Tree Grows in Paris: Lamarck's Division of Worms and Revision of Nature

I. The Making and Breaking of a Reputation

On the twenty-first day of the auspiciously named month of *Floréal* (flowering), in the spring of year 8 on the French revolutionary calendar (1800 to the rest of the Western world), the former *chevalier* (knight) but now *citoyen* (citizen) Lamarck delivered the opening lecture for his annual course on zoology at the Muséum d'Histoire Naturelle in Paris—and changed the science of biology forever by presenting the first public account of his theory of evolution. Lamarck then published this short discourse in 1801 as the first part of his treatise on invertebrate animals *(Système des animaux sans vertèbres).*

Jean-Baptiste Lamarck (1744–1829) had enjoyed a distinguished career in botany when, just short of his fiftieth birthday, he became professor of insects and worms at the Muséum d'Histoire Naturelle, newly constituted by the revolutionary government in 1793. Lamarck would later coin the term *invertebrate* for his charges. (He also introduced the word *biology* for the entire discipline in 1802.) But his original title followed Linnaeus's designation of all nonvertebrated animals as either insects or worms, a Procrustean scheme that Lamarck would soon alter. Lamarck had been an avid shell collector and student of mollusks (then classified within Linnaeus's large and heterogeneous category of Vermes, or worms)—qualifications deemed sufficient for his change of subject.

Lamarck fully repaid the confidence invested in his general biological abilities by publishing distinguished works in the taxonomy of invertebrates throughout the remainder of his career, culminating in the seven volumes of his comprehensive *Histoire naturelle des animaux sans vertèbres* (Natural history of invertebrate animals), published between 1815 and 1822. At the same time, he constantly refined and expanded his evolutionary views, extending his introductory discourse of 1800 into a first full book in 1802 (*Recherches sur l'organisation des corps vivans,* or Researches on the organization of living beings), to his magnum opus and most famous work of 1809, the two-volume *Philosophie zoologique* (Zoological philosophy), to a final statement in the long opening section, published in 1815, of his great treatise on invertebrates.

The outlines of such a career might seem to imply continuing growth of prestige, from the initial flowering to a full bloom of celebrated seniority. But Lamarck's reputation suffered a spectacular collapse, even during his own lifetime, and he died lonely, blind, and impoverished. The complex reasons for his reverses include the usual panoply of changing fashions, powerful enemies, and self-inflicted wounds based on flaws of character (in his case, primarily an overweening self-assurance that led him to ignore or underestimate the weaknesses in some of his own arguments, or the skills of his adversaries). Most prominently, his favored style of science—the construction of grand and comprehensive theories following an approach that the French call *l'esprit de système* (the spirit of system building)—became notoriously unpopular following the rise of a hard-nosed empiricist ethos in early-nineteenth-century geology and natural history.

In one of the great injustices of our conventional history, Lamarck's disfavor has persisted to our times, and most people still know him only as the foil to Darwin's greatness—as the man who invented a silly theory about giraffes stretching their necks to reach the leaves on tall trees, and then passing the fruits of their efforts to their offspring by "inheritance of acquired characters," otherwise known as the hypothesis of "use and disuse," in contrast with Darwin's proper theory of natural selection and survival of the fittest.

Indeed, the usually genial Darwin had few kind words for his French predecessor. In letters to his friends, Darwin dismissed Lamarck as an idle speculator with a nonsensical theory. In 1844, he wrote to the botanist J. D. Hooker on the dearth of evolutionary thinking (before his own ideas about natural selection): "With respect to books on the subject, I do not know of any systematical ones except Lamarck's, which is veritable rubbish." To his guru, the geologist Charles Lyell (who had accurately described Lamarck's system for English readers in the second volume of his *Principles of Geology,* published in 1832), Darwin wrote in 1859, just after publishing *The Origin of Species:* "You often allude to Lamarck's work; I do not know what you think about it, but it appeared to me extremely poor; I got not a fact or idea from it."

But these later and private statements did Lamarck no practical ill. Far more harmfully, and virtually setting an "official" judgment from that time forward, his eminent colleague Georges Cuvier—the brilliant biologist, savvy statesman, distinguished man of letters, and Lamarck's younger and antievolutionary fellow professor at the Muséum—used his established role as writer of *éloges* (obituary notices) for deceased colleagues to compose a cruel masterpiece in the genre of "damning with faint praise"—a document that fixed and destroyed Lamarck's reputation. Cuvier began with cloying praise, and then described his need to criticize as a sad necessity:

> In sketching the life of one of our most celebrated naturalists, we have conceived it to be our duty, while bestowing the commendation they deserve on the great and useful works which science owes to him, likewise to give prominence to such of his productions in which too great indulgence of a lively imagination had led to results of a more questionable kind, and to indicate, as far as we can, the cause or, if it may be so expressed, the genealogy of his deviations.

Cuvier then proceeded to downplay Lamarck's considerable contributions to anatomy and taxonomy, and to excoriate his senior colleague for fatuous speculation about the comprehensive nature of reality. He especially ridiculed Lamarck's evolutionary ideas by contrasting a caricature of his own construction with the sober approach of proper empiricism:

> These [evolutionary] principles once admitted, it will easily be perceived that nothing is wanting but time and circumstances to enable

> a monad or a polypus gradually and indifferently to transform themselves into a frog, a stork, or an elephant. . . . A system established on such foundations may amuse the imagination of a poet; a metaphysician may derive from it an entirely new series of systems; but it cannot for a moment bear the examination of anyone who has dissected a hand . . . or even a feather.

Cuvier's *éloge* reeks with exaggeration and unjust ridicule, especially toward a colleague ineluctably denied the right of response—the reason, after all, for our venerable motto, *de mortuis nil nisi bonum* ("say only good of the dead"). But Cuvier did base his disdain on a legitimate substrate, for Lamarck's writing certainly displays a tendency to grandiosity in comprehensive pronouncement, combined with frequent refusal to honor, or even to consider, alternative views with strong empirical support.

L'esprit de système, the propensity for constructing complete and overarching explanations based on general and exceptionless principles, may apply to some corners of reality, but this approach works especially poorly in the maximally complex world of natural history. Lamarck did feel drawn toward this style of system building, and he showed no eagerness to acknowledge exceptions or to change his guiding precepts. But the rigid and dogmatic Lamarck of Cuvier's caricature can only be regarded as a great injustice, for the man himself did maintain appropriate flexibility before nature's richness, and did eventually alter the central premises of his theory when his own data on the anatomy of invertebrate animals could no longer sustain his original view.

This fundamental change—from a linear to a branching system of classification for the basic groups, or phyla, of animals—has been well documented in standard sources of modern scholarship about Lamarck (principally in Richard W. Burkhardt, Jr.'s *The Spirit of System: Lamarck and Evolutionary Biology,* Harvard University Press, 1977; and Pietro Corsi's *The Age of Lamarck,* University of California Press, 1988). But the story of Lamarck's journey remains incomplete, for both the initiating incident and the final statement have been missing from the record—the beginning, because Lamarck noted his first insight as a handwritten insertion, heretofore unpublished, in his own copy of his first printed statement about evolution (the *Floréal* address of 1800, recycled as the preface to his 1801 book on invertebrate anatomy); and the ending, because his final book of 1820, *Système analytique des connaissances positives de l'homme* (Analytical system of positive knowledge about man), has been viewed only as an obscure swan song about psychology, a rare book even more rarely consulted (despite a fascinating section containing a crucial and novel wrinkle upon Lamarck's continually changing views about the classification of animals). Stories deprived of both beginnings and endings cannot satisfy our urges for fullness or completion—and I am grateful for this opportunity to supply these terminal anchors.

II. Lamarck's Theory and Our Misreadings

Lamarck's original evolutionary system—the logical, pure, and exceptionless scheme that nature's intransigent complexity later forced him to abandon—featured a division of causes into two independent sets responsible for progress and diversity respectively. (Scholars generally refer to this model as the "two-factor theory.") On the one hand, a "force that tends incessantly to complicate organization" *(la force qui tend sans cesse à composer l'organisation)* leads evolution linearly upward, beginning with spontaneous generation of "infusorians" (single-celled animals) from chemical precursors, and moving on toward human intelligence.

But Lamarck recognized that the riotous diversity of living organisms could not be ordered into a neat and simple sequence of linear advance—for what could come directly before or after such marvels of adaptation as long-necked giraffes, moles without sight, flatfishes with both eyes on one side of the body, snakes with forked tongues, or birds with webbed feet? Lamarck therefore advocated linearity only for the "principal masses," or major anatomical designs of life's basic phyla. Thus, he envisioned a linear sequence mounting, in perfect progressive regularity, from infusorian to jellyfish to worm to insect to mollusk to vertebrate. He then depicted the special adaptations of particular lineages as lateral deviations from this main sequence.

These special adaptations originate by the second set of causes, labeled by Lamarck as "the influence of circumstances" *(l'influence des circonstances).* Ironically, this second (and subsidiary) set has descended through later history as the exclusive "Lamarckism" of modern textbooks and anti-Darwinian iconoclasts (while the more important first set of linearizing forces has been forgotten). For this second set—based on change of habits as a spur to adaptation in new environmental circumstances—invokes the familiar (and false) doctrines now called "Lamarckism": the "inheritance of acquired characters" and the principle of "use and disuse."

Lamarck invented nothing original in citing these principles of inheritance, for both doctrines represented the "folk wisdom" of his time (despite their later disproof in the new world of Darwin and Mendel). Thus, the giraffe stretches its neck throughout life to reach higher leaves on acacia trees, and the shorebird extends its legs to remain above the rising waters. This sustained effort leads to longer necks or legs—and these rewards of hard work then descend to offspring in the form of altered heredity (the inheritance of acquired characters, either enhanced by use, as in these cases, or lost by disuse, as in eyeless moles or blind fishes living in perpetually dark caves).

As another irony and injustice (admittedly abetted, in part, by his own unclear statements), the ridicule that has surrounded Lamarck's theory since Cuvier's *éloge* and Darwin's dismissal has always centered upon the charge that Lamarck's views represent a sad throwback to the mystical vitalism of bad old times before modern science enshrined testable mechanical causes as the proper sources of explanation. What genuine understanding, the critics charge, can possibly arise from claims about vague and unknowable powers inherent in life itself, and propelling organisms either upward by an intrinsic complexifying force (recalling Molière's famous mock of vitalistic medicine, exemplified in the statement that morphine induces sleep *"quia est in eo virtus dormativa,"* because it contains a dormitive virtue) or sideward by some ineffable "willing" to build an adaptive branch by sheer organic effort or desire?

In a famous letter to J. D. Hooker, his closest confidant, Darwin first admitted his evolutionary beliefs in 1844 by contrasting his mechanistic account with a caricature of Lamarck's theory: "I am almost convinced . . . that species are not (it is like confessing a murder) immutable. Heaven forfend me from Lamarck nonsense of a 'tendency to progression,' 'adaptations from the slow willing of animals,' etc.! But the conclusions I am led to are not widely different from his; though the means of change are wholly so." And Cuvier, in a public forum, ridiculed the second set of adaptive forces in the same disdainful tone: "Wants and desires, produced by circumstances, will lead to other efforts, which will produce other organs. . . . It is the desire and the attempt to swim that produces membranes in the feet of aquatic birds. Wading in the water . . . has lengthened the legs of such as frequent the sides of rivers."

Lamarck hurt his own cause by careless statements easily misinterpreted in this mode. His talk about an "interior sentiment" *(sentiment intérieur)* to propel the upward force, or about organisms obeying "felt needs" (*besoins* in his terminology) to induce sideward branches of adaptation, led to suspicions about mysterious and unprovable vitalistic forces. But in fact, Lamarck remained a dedicated and vociferous materialist all his life—a credo that surely represents the most invariable and insistent claim in all his writings. He constantly sought to devise mechanical explanations, based on the physics and chemistry of matter in motion, to propel both sets of linear and lateral forces. I do not claim that his efforts were crowned with conspicuous success—particularly in his speculative attempts to explain the linear sequence of animal phyla by positing an ever more vigorous and ramifying flow of fluids, carving out spaces for organs and channels for blood in progressively more complex bodies. But one cannot deny his consistent conviction. *"La vie . . . n'est autre chose qu'un phénomène physique"* (life is nothing else than a physical phenomenon), he wrote in his last book of 1820. In a famous article, written to rehabilitate Lamarck at the Darwinian centennial celebrations (for *The Origin of Species*) in 1959, the eminent historian of science C. C. Gillispie wrote: "Life is a purely physical phenomenon in Lamarck, and it is only because science has (quite rightly) left behind his conception of the physical that he has been systematically misunderstood and assimilated to a theistic or vitalistic tradition which in fact he held in abhorrence."

Lamarck depicted his two sets of evolutionary forces as clearly distinct and destined to serve contrasting ends. The beauty of his theory—the embodiment of his *esprit de système*—lies in this clean contrast of both geometry and mechanism. The first set works upward to build progress in a strictly linear series of major anatomical designs (phyla) by recruiting a mechanism inherent in the nature of living matter. The second set works sideward to extract branches made of individual lineages (species and genera) that respond to the influence of external circumstances by precise adaptations to particular environments. (These side branches may be visualized as projecting at right angles, perpendicular to the main trunk of progress. Vectors at right angles are termed *orthogonal,* and are mathematically independent, or uncorrelated.)

Lamarck made this contrast explicit by stating that animals would form only a single line of progress if the pull of environmental adaptation did not interrupt, stymie, and divert the upward flow in particular circumstances:

> If the factor that is incessantly working toward complicating organization were the only one that had any influence on the shape and organs of animals, the growing complexity of organization would everywhere be very regular. But it is not; nature is forced to submit her works to the influence of their environment. . . . This is the special factor that occasionally produces . . . the often curious deviations that may be observed in the progression. (1809, *Philosophie zoologique;* my translations from Lamarck's original French text in all cases)

Thus, the complex order of life arises from the interplay of two forces in conflict, with progress driving lineages up the ladder, and adaptation forcing them aside into channels set by peculiarities of local environments:

> The state in which we find any animal is, on the one hand, the result of the increasing complexity of organization tending to form a regular gradation; and, on the other hand, of the influence of a multitude of very various conditions ever tending to destroy the regularity in the gradation of the increasing complexity of organization. (1809, *Philosophie zoologique*)

Finally, in all his major evolutionary works, culminating in his multivolumed treatise on invertebrate anatomy (1815), Lamarck honored the first set of linear forces as primary, and identified the second set as superposed and contrary—as in this famous statement, marking the lateral pull of adaptation as foreign, accidental, interfering, and anomalous:

> The plan followed by nature in producing animals clearly comprises a predominant prime cause. This endows animal life with the power to make organization gradually more complex. . . . Occasionally a foreign, accidental, and therefore variable cause has interfered with the execution of the plan, without, however, destroying it. This has created gaps in the series, in the form either of terminal branches that depart from the series in several points and alter its simplicity, or of anomalies observable in specific apparatuses of various organisms. (1815, *Histoire naturelle des animaux sans vertèbres*)

III. THE VALUES OF CHANGING THEORIES

Charles Darwin began the closing paragraph of his *Origin of Species* by noting the appeal of evolutionary explanation: "There is grandeur in this view of life." No thinking or feeling person can deny either nature's grandeur or the depth and dignity of our discovery that a history of evolution binds all living creatures together. But in our world of diverse passions and psychologies, primary definitions (and visceral feelings) about grandeur differ widely among students of natural history. Darwin emphasized the bounteous diversity itself, in all its buzzing and blooming variety—for the last sentence of this closing paragraph contrasts the "dullness" of repetitive planetary cycling with the endless expansion and novelty of evolution's good work: "whilst this planet has gone cycling on according to the fixed law of gravity, from so simple a beginning endless forms most beautiful and most wonderful have been, and are being, evolved."

But I suspect that Lamarck, following his own upbringing in the rigorous traditions of French rationalism during the Enlightenment, construed the primary definition of grandeur quite differently. As a devotee of *l'esprit de système,* Lamarck surely viewed the capacity of the human mind (his own in this case, for he was not a modest man) to apprehend the true and complete system of nature's rational order as a primary criterion (with the actual production of riotous diversity as a consequence requiring taxonomic arrangement, but only a product of the less important lateral set of adaptive forces that disturb the march of progress). Thus, the logical clarity of his two-factor theory—with a primary cause establishing a linear march of rational progress and an opposed and subsidiary cause generating a more chaotic forest of adaptive diversity—must have struck Lamarck as the defining ingredient of nature's grandeur and the power of evolution.

Our understanding of nature must always reflect a subtle interaction between messages from genuine phenomena truly "out there" in the real world and the necessary filtering of such data through all the foibles and ordering devices internal to the human mind and its evolved modes of action (see chapter 2). We cannot comprehend nature's complexity—particularly for such comprehensive subjects as evolution and the taxonomic structure of organic diversity—unless we impose our mental theories of order upon the overt chaos that greets our senses. The different styles followed by scientists to balance and reconcile these two interacting (but partly contradictory) sources of order virtually define the rich diversity of fruitful approaches pursued by a profession too often, and falsely, caricatured as a monolithic enterprise committed to a set of fixed procedures called "*the* scientific method."

Dangers and opportunities attend an overemphasis on either side. Rigid systematizers often misconstrue natural patterns by forcing their observations into rigidly preconceived structures of explanation. But colleagues who try to approach nature on her own terms, without preferred hypotheses to test, risk either being overwhelmed by a deluge of confusing information or falling prey to biases that become all the more controlling by their unconscious (and therefore unrecognized) status.

In this spectrum of useful approaches, Lamarck surely falls into the domain of scientists who place the logical beauty of a fully coherent theory above the messiness of nature's inevitable nuances and exceptions. In this context, I am all the more intrigued by Lamarck's later intellectual journey, so clearly contrary to his own inclinations, and inspired (in large part) by his inability to encompass new discoveries about the anatomy of invertebrates into the rigid confines of his beautiful system.

Nothing in the history of science can be more interesting or instructive than the intellectual drama of such a slow transformation in a fundamental view of life—from an initial recognition of trouble, to attempts at accommodation within a preferred system, to varying degrees of openness toward substantial change, and sometimes, among the most flexible and courageous, even to full conversion. I particularly like to contemplate the contributions of external and internal factors to such a change: new data mounting a challenge from the outside, coordinated with an internal willingness to follow the logic of an old system to its points of failure, and then to construct a revised theory imposing a different kind of consistency upon an altered world (with minimal changes for those who remain in love with their previous certainties and tend to follow conservative intellectual strategies, or with potentially revolutionary impact for people with temperaments that permit, or even favor, iconoclasm and adventure). Reward and risk go hand in hand, for the great majority of thoroughly radical revisions must fail, even though the sweetest fruits await the few victors in this chanciest and most difficult of all mental adventures.

When we can enjoy the privilege of watching a truly great intellect struggling with the most important of all biological concepts at a particularly interesting time in the history of science, then all factors coincide to produce a wonderful story offering unusual insight into the workings of science as well. When we can also experience the good fortune of locating a previously missing piece—in this case, the first record of a revision that would eventually alter the core of a central theory, although Lamarck, at this inception, surely had no inkling of how vigorously such a small seed could grow—then we gain the further blessing of an intriguing particular (the substrate of all good gossip) grafted onto a defining generality. The prospect of being an unknown witness—the "fly on the wall" of our usual metaphor—has always excited our fancy. And the opportunity to intrude upon a previously undocumented beginning—to be "present at the creation" in another common description—evokes an extra measure of intrigue. In this case, we begin with something almost inexpressibly humble: the classification of worms—and end with both a new geometry for animal life and a revised view of evolution itself.

IV. Lamarck Emends His First Evolutionary Treatise

Once upon a time, in a faraway world before the electronic revolution, and even before the invention of typewriters, authors submitted literal manuscripts (from the Latin for "written by hand") to their publishers. When scholars revised a book for a second edition, they often worked from a specially prepared "interleaved copy" containing a blank sheet after each printed page. Corrections and additions could then be written on the blanks, enabling publishers to set a new edition from a coherent, bound document (rather than from a confusing mess of loose or pasted insertions).

Lamarck owned an interleaved copy of his first evolutionary treatise—*Système des animaux sans vertèbres* of 1801. Although he never published a second edition, he did write comments on the blank pages—and he incorporated some of these statements into later works, particularly *Philosophie zoologique* of 1809. This copy, which might have tempted me to a Faustian form of collusion with Mephistopheles, recently sold at auction for more money than even a tolerably solvent professor could ever dream of having at his disposition. But I was able to play the intellectual's usual role of voyeur during the few days of previewing before the sale—and I did recognize, in a crucial note in Lamarck's hand, a significance that had eluded previous observers. The eventual buyer (still unknown to me) expressed gratitude for the enhanced importance of his purchase, and kindly offered (through the bookseller who had acted as his agent) to lend me the volume for a few days, and to allow me to publish the key note in this forum. I floated on cloud nine and happily rooted like a pig in . . . during those lovely days when I could hold and study my profession's closest approach to the holy grail.

Lamarck did not make copious additions, but several of his notes offer important insights, while their general tenor teaches us something important about the relative weighting of his concerns. The first forty-eight pages of the printed book contain the *Floréal* address, Lamarck's initial statement of his evolutionary theory. The final 350 pages present a systematic classification of invertebrates, including a discussion of principles and a list and description of all recognized genera, discussed phylum by phylum.

Of Lamarck's thirty-seven handwritten additions on the blank pages, twenty-nine offer only a word or two, and represent the ordinary activity of correcting small errors, inserting new information, or editing language. Lamarck makes fifteen comments about anatomy, mostly in his chapter on the genera of mollusks, the group he knew best. A further nine comments treat taxonomic issues of naming (adding a layman's moniker to the formal Latin designation, changing the name or affiliation of a genus); two add bibliographic data; and the final three edit some awkward language.

Taken as an ensemble, I regard these comments as informative in correcting a false impression that Lamarck, by this time of seniority in his career, cared only for general theory, and not for empirical detail. Clearly, Lamarck continued to cherish the minutiae of raw information, and to keep up with developing knowledge—the primary signs of an active scientific life.

Of the eight longer comments, four appear as additions to the *Floréal* address. They provide instructive insight into Lamarck's character and concerns by fulfilling the "conservative" function of making more explicit, and elaborating by hypothetical examples, the central feature of his original evolutionary theory: the sharp distinction between causes of "upward" progress and "sideward" adaptation to local circumstances.

In the two comments (among these four) that attracted most attention from potential buyers, Lamarck added examples of adaptation to local environments by inheritance of acquired characters (with both cases based on the evolutionary reorientation of eyes): first, flatfishes that live in shallow water, flatten their bodies to swim on their side, and then move both eyes to the upper surface of the head; and second, snakes that move their eyes to the top of their head because they live so close to the ground and must therefore be able to perceive a world of danger above them—and then need to develop a long and sensitive tongue to perceive trouble in front, and now invisible to the newly placed eyes. With these examples, Lamarck generalized his second set of forces by extending his stories to a variety of organisms: the *Floréal* address had confined all examples to the habits and anatomy of birds. (The purely speculative character of these cases also helps us to understand why more sober empiricists, like Darwin and Cuvier, felt so uncomfortable with Lamarck's supposed data for evolution.) In any case, Lamarck published both examples almost verbatim in his *Philosophie zoologique* of 1809.

A third comment then strengthens the other, and primary, set of linear causes by arguing that the newly discovered platypus of Australia could link the penultimate birds to the highest group of mammals. Finally, the fourth comment tries to explain the mechanisms of use and disuse by differential flow of fluids through bodies.

The other set of four longer comments adorns the second part of the book on taxonomic ordering of invertebrate animals. One insertion suggests that a small and enigmatic egg-shaped fossil should be classified within the phylum of corals and jellyfishes. A second statement, of particular interest to me, revises Lamarck's description of the clam genus *Trigonia*. This distinctive form had long been recognized as a prominent fossil in Mesozoic rocks, but no Tertiary fossils or living specimens had ever been found—and naturalists therefore supposed that this genus had become extinct. But two French naturalists then

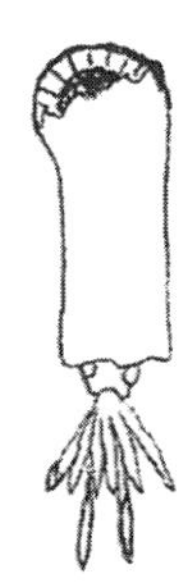

Lamarck's original illustration depicting the discovery of the squidlike animal that secretes the shell called Spirula.

céphalopode [calmar] ayant une
coquille enchassée dans la partie ou l'ex-
trémité postérieure de son Corps.

found a living species of *Trigonia* in Australian waters, and Lamarck himself published the first description of this triumphant rediscovery in 1803. (As an undergraduate, I did my first technical research on dissection, and also wrote my first paper in the history of science, under the direction of Norman D. Newell at the Museum of Natural History in New York. He gave me a half dozen, still preciously rare, specimens of modern Australian trigonians. When I gulped and admitted that I had no experience with dissection and feared butchering such a valuable bounty, he said to me, in his laconic manner—so inspirational for self-motivated students, but so terrifying for the insecure—"Go down to the Fulton Fish Market and buy a bunch of quahogs. Practice on them first." I was far more terrified than inspired, but all's well that ends well.)

The final two comments provide the greatest visceral pleasure of all because Lamarck added drawings to his words (reproduced here with the kind permission of the book's new owner). The first sketch affirms Lamarck's continuing commitment to detail, and to following and recording new discoveries. A small, white, and delicate coiled shell of a cephalopod mollusk (the group including squid and octopuses) frequently washes up on beaches throughout the world. Lamarck himself had named this shell *Spirula* in 1799. But the animal that makes the shell had never been found. As a particular mystery, no one knew whether the animal lived inside the shell (as in a modern chambered nautilus) or grew the shell within its body (as in the "cuttlebone" of a modern squid). The delicacy of the object suggested a protected internal status, but the question remained open. Soon after Lamarck's book appeared, naturalists discovered the animal of *Spirula,* and affirmed an internal shell—a happy resolution that inspired Lamarck to a rare episode of artistic activity.

The last—and, as I here suggest, by far the most important—comment appears on the blank sheet following page 330, which contains the description of two remarkably different genera of "worms"—the medicinal leech *Hirudo* and the pond worm *Planaria,* known to nearly anyone who ever took a basic

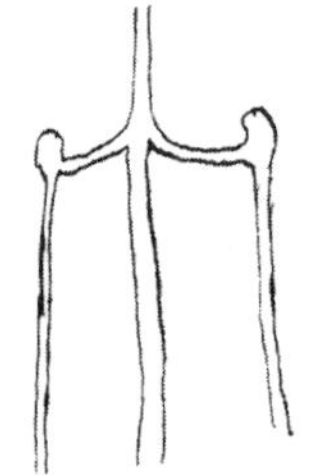

dans les vers annelés et qui ont des organs externes,
le sang est rouge et circule dans des vaisseaux arte-
riels et veineux. leur organisation les place avant
les insectes.
les vers intestins doivent seuls se trouver après les
insectes. ils n'ont qu'un fluide blanc, libre, non
contenu dans des vaisseaux. cuvier. extrait d'un
mém. lu à l'institut le 11 nivose an 10.

Lamarck's original drawing and text, expressing his first and crucial recognition that annelid worms and parasitic internal worms represent very different kinds of animals.

---

laboratory course in biology. Here, Lamarck draws a simple sketch of the circulatory system of an annelid worm, and then writes the following portentous words:

> *observation sur l'org^{on} des vers. dans les vers annelés et qui ont des organs externes, le sang est rouge et circule dans des vaisseaux arteriels et veineux. leur organisation les place avant les insectes. les vers intestins doivent seuls se trouver après les insectes. ils n'ont qu'un fluide blanc, libre, non contenue dans des vaisseaux. Cuvier. extrait d'un mém. lu a l'institut le 11 nivôse an 10.*
>
> (Observation on the organization of worms. In annelid worms, which have external organs, the blood is red and circulates in arterial and venous vessels. Their organization places them before the insects. Only the internal worms come after the insects. They have only a white fluid, free, and not contained in vessels. Cuvier. Extract from a memoir read at the institute on the 11th day of [the month of] Nivôse year 10.)

Clearly, Lamarck now recognizes a vital distinction between two groups previously lumped together into the general category of "worms." He regards one group—the annelids, including earthworms, leeches, and the marine polychaetes—as highly advanced, even more so than insects. (Lamarck usually presented his scale of animal life from the top down, starting with humans and ending with infusorians—and not from the bottom up, the system that became conventional in later zoological writing. Thus, he states that annelids come *before* insects because he views them as more advanced—that is, closer to the mammalian top.) But another group, the internal worms* (mostly parasites living within the bodies of other animals), rank far lower on the scale—even after (that is, anatomically simpler than) insects. These two distinct groups, previously conflated, must now be widely separated in the taxonomic ordering of life. Ironically, Lamarck acknowledges his colleague Cuvier (who would later turn against him and virtually destroy his reputation) as the source for a key item of information that changed his mind—Cuvier's report (presented at a meeting during the winter of 1801–2, soon after the publication of Lamarck's book) that annelids possessed a complicated circulatory system, with red blood running in arteries and veins, whereas internal worms grew no discrete blood vessels, and only moved a white fluid through their body cavity.

Obviously, Lamarck viewed this new information as especially important, for no other anatomical note received nearly such prominence in his additions, while only one other observation (a simple new bit of information, without much theoretical meaning) merits a drawing. But why did Lamarck view this division of worms as so important? And how could such an apparently dull and technical decision about naming act as a pivot and initiator for a new view of life?

V. An Odyssey of Worms

I have always considered it odd (and redolent either of arrogance or parochiality) when a small minority divides the world into two wildly unbalanced categories of itself versus all others—and then defines the large category as an absence of the small, as in my grandmother's taxonomy for *Homo sapiens:* Jews and non-Jews. Yet our conventional classification of animals follows the same strategy by drawing a basic distinction between vertebrates and invertebrates—when only about forty thousand of more than a million named species belong to the relatively small lineage of vertebrates.

*The standard English literature on this subject always translates Lamarck's phrase incorrectly as "intestinal worms." These parasites dwell in several organs and places of vertebrate (and other) bodies, not only in the intestines. In French, the word *intestin* conveys the more general meaning of "internal" or "inside."

On the venerable principle that bad situations can always be made worse, we can gain some solace by noting the even greater imbalance devised by the founder of modern taxonomy, Carolus Linnaeus. At least we now recognize vertebrates as only part of a single phylum, while most modern schemes divide invertebrates into some twenty to thirty separate phyla. But in his *Systema naturae* of 1758, the founding document of modern zoological nomenclature, Linnaeus identified only six basic animal groups: four among vertebrates (mammals, birds, reptiles, and fishes), and two for the entire realm of invertebrates (Insecta, for insects and their relatives, and Vermes, literally worms, for nearly everything else).

When Lamarck became professor of invertebrates at the Muséum in 1793 (with an official title in a Linnaean straitjacket as professor of insects and worms), he already recognized that reform demanded the dismemberment of Linnaeus's "wastebucket" category of Vermes, or worms. (*Wastebucket,* by the way, actually ranks as a semitechnical term among professional taxonomists, a description for inflated groups that become receptacles for heterogeneous bits and pieces that most folks would rather ignore—as in this relegation of all "primitive" bilaterally symmetrical animals to a category of "worms" ranking far beneath the notice of specialists on vertebrates.)

In his 1801 book, Lamarck identified the hodgepodge of Linnaeus's Vermes as the biggest headache and impediment in zoology:

> The celebrated Linnaeus, and almost all other naturalists up to now, have divided the entire series of invertebrate animals into only two classes: insects and worms. As a consequence, anything that could not be called an insect must belong, without exception, to the class of worms.

By the time Lamarck wrote his most famous book in 1809, his frustration had only increased, as he called Linnaeus's class of worms *"une espèce de chaos dans lequel les objets très-disparates se trouvent réunis"* (a kind of chaos where very disparate objects have been united together). He then blamed the great man himself for this sorry situation: "The authority of this scientist carried such great weight among naturalists that no one dared to change this monstrous class of worms." (I am confident that, in writing *"cette classe monstrueuse,"* Lamarck meant to attack the physical size based on number of included genera, not the moral status, of Linnaeus's Vermes.)

Lamarck therefore began his campaign of reform by raiding Vermes and gradually adding the extracted groups as novel phyla in his newly named category of invertebrates. In his first course of 1793, he had already expanded the Linnaean duality to a ladder of progress with five rungs—mollusks, insects, worms, echinoderms, and polyps (corals and jellyfish)—by liberating three new phyla from the wastebucket of Vermes.

This reform accelerated in 1795, when Georges Cuvier arrived and began to study invertebrates as well. The two men collaborated in friendship at first—and they surely operated as one mind on the key issue of dismembering Vermes. Thus, Lamarck continued to add phyla in almost every annual course of lectures, extracting most new groups from Vermes, but some from the overblown Linnaean Insecta as well. In year 7 (1799), he established the Crustacea (for marine arthropods, including crabs, shrimp, and lobsters), and in year 8 (1800) the Arachnida (for spiders and scorpions). Lamarck's invertebrate classification of 1801 therefore featured a growing ladder of progress, now bearing seven rungs. In 1809, he presented a purely linear sequence of progress for the last time in his most famous book, *Philosophie zoologique.* His tall and rigid ladder now included fourteen rungs, as he added the four traditional groups of vertebrates atop a list of invertebrate phyla that had just reached double digits (see accompanying chart directly reproduced from the 1809 edition).

So far, Lamarck had done nothing to inspire any reconsideration of the evolutionary views first presented in his *Floréal* address of 1800. His taxonomic reforms, in this sense, had been entirely conventional in adding weight and strength to his original views. The *Floréal* statement had contrasted a linear force leading to progress in major groups with a lateral force causing local adaptation in particular lineages. Lamarck's ladder included only seven groups in the *Floréal*

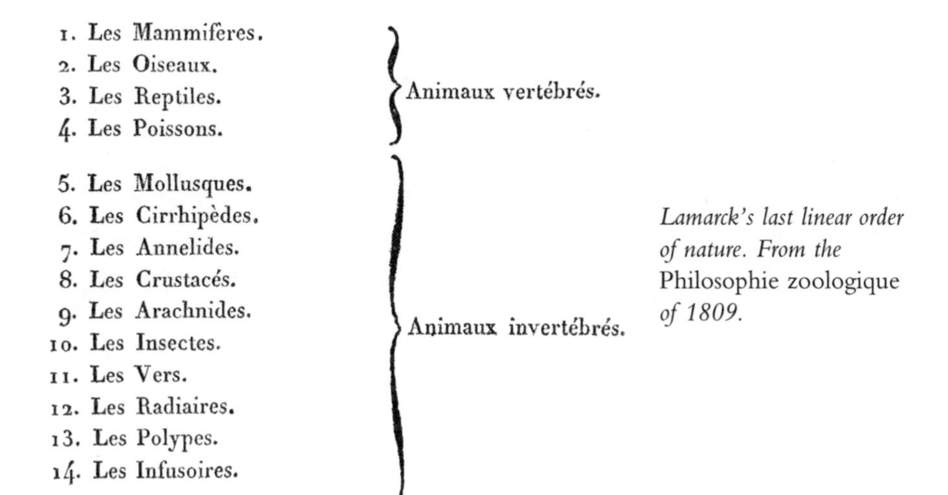
1. Les Mammifères.
2. Les Oiseaux.
3. Les Reptiles.
4. Les Poissons.

} Animaux vertébrés.

5. Les Mollusques.
6. Les Cirrhipèdes.
7. Les Annelides.
8. Les Crustacés.
9. Les Arachnides.
10. Les Insectes.
11. Les Vers.
12. Les Radiaires.
13. Les Polypes.
14. Les Infusoires.

} Animaux invertébrés.

Lamarck's last linear order of nature. From the Philosophie zoologique *of 1809.*

address. By 1809, he had doubled the length while preserving the same strictly linear form—thus strengthening his central contrast between two forces by granting the linear impetus a greatly expanded field for its inexorably exceptionless operation.

But if Lamarck's first reform of Linnaeus—the expansion of groups into a longer linear series—had conserved and strengthened his original concept of evolution, he now embarked upon a second reform, destined (though he surely had no inkling at the outset) to yield the opposite effect of forcing a fundamental change in his view of life. He had, heretofore, only extracted misaligned groups from Linnaeus's original Vermes. He now needed to consider the core of Vermes itself—and to determine whether waste and rot existed at the foundation as well.

"Worms," in our vernacular understanding, are defined both broadly and negatively (unfortunate criteria guaranteeing inevitable trouble down the road) as soft-bodied, bilaterally symmetrical animals, roughly cylindrical in shape and lacking appendages or prominent sense organs. By these criteria, both earthworms and tapeworms fill the bill. For nearly ten years, Lamarck did not seriously challenge this core definition.

But he could not permanently ignore the glaring problem, recognized but usually swept under the rug by naturalists, that this broad vernacular category seemed to include at least two kinds of organisms bearing little relationship beyond a superficial and overt similarity of external form. On the one hand, a prominent group of free-living creatures—earthworms and their allies—built bodies composed of rings or segments, and also developed internal organs of substantial complexity, including nerve tubes, blood vessels, and a digestive tract. But another assemblage of largely parasitic creatures—tapeworms and their allies—grew virtually no discretely recognizable internal organs at all, and therefore seemed much "lower" than earthworms and their kin under any concept of an organic scale of complexity. Would the heart of Vermes therefore need to be dismembered as well?

This problem had already been worrying Lamarck when he published the *Floréal* address in his 1801 compendium on invertebrate anatomy—but he was not yet ready to impose a formal divorce upon the two basic groups of "worms." Either standard of definition, taken by itself—different anatomies or disparate environments—might not offer sufficient impetus for thoughts about taxonomic separation. But the two criteria conspired perfectly together in the remaining Vermes: the earthworm group possessed complex anatomy *and* lived freely in the outside world; the tapeworm group maintained maximal simplicity among mobile animals *and* lived almost exclusively within the bodies of other creatures.

Lamarck therefore opted for an intermediary solution. He would not yet dismember Vermes, but he would establish two subdivisions *within* the class: *vers externes* (external worms) for earthworms and their allies, and *vers intestins* (internal worms) for tapeworms and their relatives. He stressed the simple anatomy of the parasitic subgroup, and defended their new name as a spur to further study, while arguing that knowledge remained insufficient to advocate a deeper separation:

> It is very important to know them [the internal worms], and this name will facilitate their study. But aside from this motive, I also believe that such a division is the most natural . . . because the internal worms are much more imperfect and simply organized than the other worms. Nevertheless, we know so little about their origin that we cannot yet make them a separate order.

At this point, the crucial incident occurred that sparked Lamarck to an irrevocable and cascading reassessment of his evolutionary views. He attended Cuvier's lecture during the winter of 1801–2 (year 10 of the revolutionary calendar), and became convinced, by his colleague's elegant data on the anatomy of external worms, that the extensive anatomical differences between his two subdivisions could not permit their continued residence in the same class. He would, after all, have to split the heart of Vermes. Therefore, in his next course, in the spring of 1802, Lamarck formally established the class Annelida for the external worms (retaining Vermes for the internal worms alone), and then separated the two classes widely by placing his new annelids above insects in linear complexity, while leaving the internal worms near the bottom of the ladder, well below insects.

Lamarck formally acknowledged Cuvier's spur when he wrote a history of his successive changes in classifying invertebrates for the *Philosophie zoologique* of 1809:

> Mr. Cuvier discovered the existence of arterial and venous vessels in distinct animals that had been confounded with other very differently organized animals under the name of worms. I soon used this new fact to perfect my classification; therefore, I established the class of annelids in my course for year 10 (1802).

The handwritten note and drawing in Lamarck's 1801 book, discussed and reproduced earlier in this essay, tells much the same story—but what a contrast, in both intellectual and emotional intrigue, between a sober memory written long after an inspiration, and the inky evidence of the moment of enlightenment itself!

But this tale should now be raising a puzzle in the minds of readers. Why am I making such a fuss about this particular taxonomic change—the final division of Vermes into a highly ranked group of annelids and a primitive class of internal worms? In what way does this alteration differ from any other previously discussed? In all cases, Lamarck subdivided Linnaeus's class Vermes and established new phyla in his favored linear series—thus reinforcing his view of evolution as built by contrasting forces of linear progress and lateral adaptation. Wasn't he just following the same procedure in extracting annelids and placing them on a new rung of his ladder? So it might seem—at first. But Lamarck was too smart, and too honorable, to ignore a logical problem directly and inevitably instigated by this particular division of worms—and the proper solution broke his system.

At first, Lamarck did treat the extraction of annelids as just another addition to his constantly improving linear series. But as the years passed, he became more and more bothered by an acute problem, evoked by an inherent conflict between this particular taxonomic decision and the precise logic of his overarching system. Lamarck had ranked the phylum Vermes, now restricted to the internal worms alone, just above a group that he named *radiaires*—actually (by modern understanding) a false amalgam of jellyfishes from the coelenterate phylum and sea urchins and their relatives from the echinoderm phylum. Worms had to rank above radiates because bilateral symmetry and directional motion trump radial symmetry and an attached (or not very mobile) lifestyle—at least in conventional views about ladders of progress (which, of course, use mobile and bilaterally symmetrical humans as an ultimate standard). But the parasitic internal worms also lack the two most important organ systems—nerve ganglia and cords, and circulatory vessels—that virtually define complexity on the traditional ladder. Yet echinoderms within the "lower" radiate phylum develop both nervous and circulatory systems. (These organisms circulate sea water rather than blood, but they do run their fluids through tubes.)

If the primary "force that tends incessantly to complicate organization" truly works in a universal and exceptionless manner, then how can such an inconsistent situation arise? If the force be general, then any given group must stand fully higher or lower than any other. A group cannot be higher for some features, but lower for others. Taxonomic experts cannot pick and choose. He who lives by the line must die by the line.

This problem did not arise so long as annelids remained in the class of worms. Lamarck, after all, had never argued that each genus of a higher group must rank above all members of a lower group in every bodily part. He only claimed that the "principal masses" of organic design must run in pure linear order. Individual genera may degenerate or adapt to less complex environments in various parts—but so long as some genera display the higher conformation in all features, then the entire group retains its status. In this case, so long as annelids remained in the group, then many worms possessed organ systems more complex than any comparable part in any lower group—and the entire class of worms could retain its unambiguous position above radiates and other primitive forms. But with the division of worms and the banishment of complex annelids, Lamarck now faced the logical dilemma of a coherent group (the internal parasitic worms alone) higher than radiates in some key features but lower in others. The pure march of nature's progress—the keystone of Lamarck's entire system—had been fractured.

Lamarck struggled with this problem for several years. He stuck to the line of progress in 1802, and again—for the last time, and in a particularly uncompromising manner that must, in retrospect, represent a last hurrah before the fall—in the first volume of his seminal work, *Philosophie zoologique,* of 1809. But honesty eventually trumped hope. Just before publication, Lamarck appended a short chapter of "additions" to volume two of *Philosophie zoologique.* He now, if only tentatively, floated a new scheme that would resolve his problem with worms, but would also unravel his precious linear system.

Lamarck had long argued that life began with the spontaneous generation of "infusorians" (single-celled animals) in ponds. But suppose that spontaneous generation occurs twice, and in two distinct environments—in the external world for a lineage beginning with infusorians, and inside the bodies of other creatures for a second lineage beginning with internal worms? Lamarck therefore wrote that "worms seem to form one initial lineage in the scale of animals, just as, evidently, the infusorians form the other branch."

Lamarck then faced the problem of allocating the higher groups. To which of the two great lines does each belong? He presented his preliminary thoughts in a chart—perhaps the first evolutionary branching diagram ever published in the history of biology—that directly contradicted his previous image of a single ladder. (Compare this figure with the version presented earlier in this essay, taken from volume one of the same 1809 work.) Lamarck begins (at the top, contrary to current conventions) with two lines, labeled *"infusoires"* (single-celled animals) and *"vers"* (worms). He then inserts light dots to suggest possible allocations of the higher phyla to the two lines. The logical problem that broke his system has now been solved—for the *radiaires* (radiate animals), standing below worms in some features, but above in others, now rank in an entirely separate series, directly following an infusorian beginning.

When mental floodgates open, the tide of reform must sweep to other locales. Once he had admitted branching and separation at all, Lamarck could hardly avoid the temptation to apply this new scheme to other old problems. Therefore, he also suggested some substantial branching at the end of his array. He had always been bothered by the conventional summit of reptiles to birds to mammals, for birds seemed just different from, rather than inferior to, mammals. Lamarck therefore proposed (and drew on his revolutionary chart) that reptiles branched at the end of the series, one line passing from turtles to birds *(oiseaux)* to *monotrèmes* (platypuses, which Lamarck now considers as separate from mammals), the other from crocodiles to marine mammals (labeled *m. amphibies*) to terrestrial mammals. Finally, and still in the new spirit, he even posited a threefold branching in the transition to terrestrial mammals, leading to separate lines for whales *(m. cétacés),* hoofed animals *(m. ongulés),* and mammals with nails *(m. onguiculés),* including carnivores, rodents, and primates (including humans).

Finally, Lamarck explicitly connected the two reforms: the admission of two sequences of spontaneous generation at the bottom, and a branching among higher vertebrates at the top: "The animal scale begins with at least two branches; in the course of its extent, several branches seem to end in different places."

After *Philosophie zoologique* of 1809, Lamarck wrote one additional major book on evolution, the introductory volume (1815) to his *Histoire naturelle des animaux sans vertèbres.* Here, he abandoned all the tentativeness of his 1809 revision, and announced his conversion to branching as the fundamental pattern of evolution. In direct contradiction to the linear model that had shaped all his previous work, Lamarck stated simply, and without ambiguity:

> *Dans sa production des differents animaux, la nature n'a pas executé une série unique et simple.*
> [In its production of the different animals, nature has not fashioned a single and simple series.]

He then emphasized the branching form of his new model, and explained how the division of worms, inspired by Cuvier's observations, had broken his former system and impelled his revision:

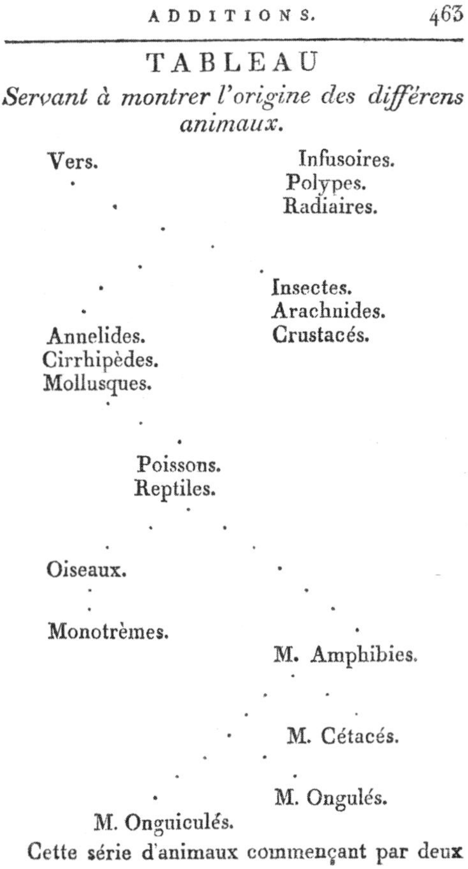

ADDITIONS. 463

TABLEAU

Servant à montrer l'origine des différens animaux.

Vers.

Infusoires.
Polypes.
Radiaires.

Insectes.
Arachnides.
Crustacés.

Annelides.
Cirrhipèdes.
Mollusques.

Poissons.
Reptiles.

Oiseaux.

Monotrèmes.

M. Amphibies.

M. Cétacés.

M. Ongulés.

M. Onguiculés.

Cette série d'animaux commençant par deux

Lamarck's first depiction of a branching model for the history of life. From the appendix to Philosophie zoologique *(1809).*

> The order is far from being simple; it is branching [*rameux*] and even appears to be constructed of several distinct series. . . . The animals that belonged to the class of worms display a great disparity of organization. . . . The most imperfect of these animals arise by spontaneous generation, and the worms [now restricted to *vers intestins,* with annelids removed] truly form their own series, later in origin than the one that began with infusorians.

Lamarck's third and last chart (reproduced here from his 1815 volume) shows how far he had progressed both in his own confidence, and in copious branching on his new tree of life. He titles the chart "presumed order of the formation of animals, showing two separate and subbranching series." Note how the two major lines of separate spontaneous generation—one beginning with infusorians, the other with internal worms—are now clearly marked and separated. Note also how each of the series also divides within itself, thus establishing the process of branching as a key theme at all scales of the system. The infusorian line branches at the level of polyps (corals and jellyfish) into a line of radiates and a line terminating in mollusks. The second line of worms also branches in two, leading to annelids on one side and insects on the other. But the insect line then splits again (a tertiary division) into a lineage of crustaceans and barnacles (labeled *cirrhipèdes*) and another of arachnids (spiders and scorpions).

Finally, we must recognize that these major changes do not only affect the overt geometry of animal organization. The conversion from linearity to branching also—perhaps even more importantly—marks a profound shift in Lamarck's underlying theory of nature. He had based his original system, defended explicitly and vociferously until 1809, on a fundamental division of two independent forces—a primary cause that builds basic anatomies in an unbroken line of progress, and a subsidiary lateral force that draws single lineages off the line into byways of immediate adaptation to local environments. A set of philosophical consequences then spring from this model: the predictable and lawlike character of evolution lies patent in the primary force and its ladder of progress reaching to man, while accidents of history (leading to local adaptations) can then be dismissed as secondary and truly independent from the overarching order.

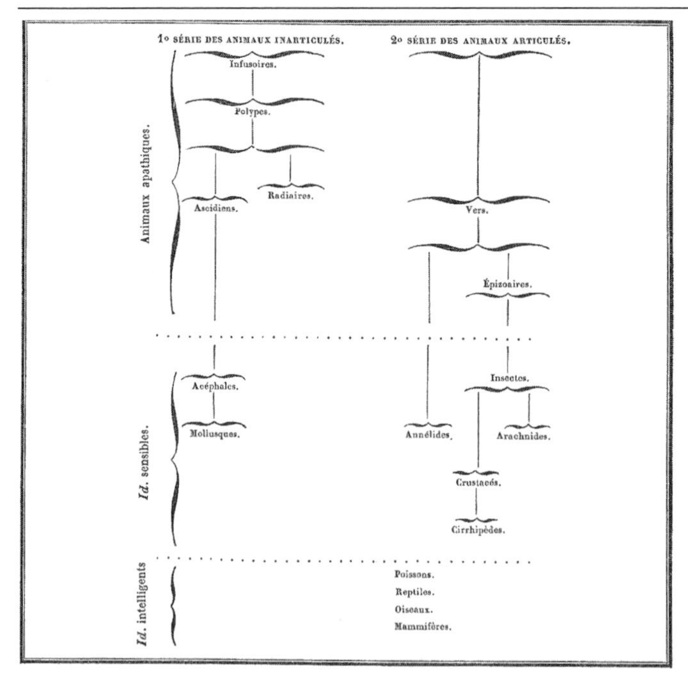

Lamarck's fully developed tree of life from 1815.

But the branching system destroys this neat and comforting scheme. First of all, the two forces become intermingled and conflated in the branching itself. We can no longer distinguish two independent and orthogonal powers working at right angles. Progress may occur along any branch to be sure, but the very act of division implies an environmental impetus to split the main line—and Lamarck had always advocated a complete and principled distinction between a single and inexorable main line and the numerous minor deviations that can draw off a long-necked giraffe or an eyeless mole, but can never disrupt or ramify the major designs of animal life. In the new model, however, environment intrudes at the first construction of basic order—as one group arises spontaneously in ponds, and another inside the bodies of other creatures! Moreover, each of the two resulting lines then branches further, and unpredictably, under environmental impetuses that were not supposed to derail the force of progress among major groups—as when insects split into a terrestrial line of arachnids and a marine line leading to crustaceans and barnacles.

Second, the forces of history and natural complexity have now triumphed over the scientific ideal of a predictable and lawlike system. The taxonomy of animals could no longer embody an overarching plan of progress, illustrating the fundamental order, harmony, and predictable good sense of the natural world (perhaps even the explicit care of a loving deity, whose plans we may hope to understand because he thinks as we do). Now, the confusing, particular, local, and unpredictable forces of complex environments hold sway, ready at any time to impose a deviation upon any group with enough hubris to suppose that Emerson's forthcoming words could describe their inevitable progress:

And striving to be man, the worm
Mounts through all the spires of form.

VI. Lamarck's Epilogue and My Own

Following his last and greatest treatise on the anatomy of invertebrate organisms, Lamarck published only one other major work—*Analytic System of Positive Knowledge About Man* (1820). This rare book has not been consulted by previous historians who traced the development of Lamarck's changing views on the classification of animals. Thus, traditional accounts stop at Lamarck's 1815 revision, with its fundamental distinction between two separate lineages of spontaneous generation. The impression therefore persists that Lamarck never fully embraced the branching model, later exemplified by Darwin as the "tree of life"—with a common trunk of origin for all creatures and no main line of growth thereafter. Lamarck had compromised his original ladder of progress by advocating two separate origins, but he could continue to stress linearity in each of the resulting series.

But his 1820 book, although primarily a treatise on psychology, does include a chapter on the classification of animals—and I discovered, in reading these pages, that Lamarck did pursue his revisionary path further, and did finally arrive at a truly branching model for a tree of life. Moreover, in a remarkable passage, Lamarck also recognizes the philosophical implications of his full switch by acknowledging a reversal in his ranking of natural forces in one of the most interesting (and honorable) intellectual conversions that I have ever read.

Lamarck still talks about forces of progress and forces of branching, and he does argue that progress will proceed along each branch. But branching has triumphed as a primary and controlling theme, and Lamarck now frames his entire discussion of animal taxonomy by emphasizing successive points of division. For example, consider this epitome of vertebrate evolution:

> Reptiles come necessarily after fishes. They build a branching sequence, with one branch leading from turtles to platypuses to the diverse group of birds, while the other seems to direct itself, via lizards, toward the mammals. The birds then . . . build a richly varied branching series, with one branch ending in birds of prey.

(In previous models, Lamarck had viewed birds of prey as the top rung of a single avian ladder.)

But much more radically, his 1815 model based on two lines of spontaneous generation has now disappeared. In its place, Lamarck advocates the same tree of life that would later become conventional through the influence of Darwin and other early evolutionists. Lamarck now proposes a single common ancestor for all animals, called a monad. From this beginning, infusorians evolve, followed by polyps, arising "directly and almost without a gap." But polyps then branch to build the rest of life's tree: "instead of continuing as a single series, the polyps appear to divide themselves into three branches"—the radiates, which end without evolving any further; the worms, which continue to branch into all phyla of segmented animals, including annelids, insects, arachnids, crustaceans, and barnacles, each by a separate event of division; and the tunicates (now regarded as marine organisms closely related to vertebrates), which later split to form several lines of mollusks and vertebrates.

Lamarck then acknowledges the profound philosophical revision implied by a branching model for nature's fundamental order. He had always viewed the linear force of progress as primary. As late as 1815, even after he had changed his model to permit extensive branching and two environmentally induced sequences of spontaneous generation, Lamarck continued to emphasize the primary power of the linear force, compared with disturbing and anomalous exceptions produced by lateral environmental causes, called *l'influence des circonstances.* To restate the key passage quoted earlier in this essay:

> The plan followed by nature in producing animals clearly comprises a predominant prime cause. This endows animal life with the power to make organization gradually more complex. . . . Occasionally a foreign, accidental, and therefore variable cause has interfered with the execution of the plan . . . [producing] branches that depart from the series in several points and alter its simplicity.

But Lamarck, five years later in his final book of 1820, now abandons this controlling concept of his career, and embraces the opposite conclusion. The influence of circumstances (leading to a branching model of animal taxonomy) rules the paths of evolution. All general laws, of progress or anything else, must be regarded as subservient to the immediate singularities of environments and histories. The influence of circumstances has risen from a disturbing and peripheral joker to the true lord of all (with an empire to boot):

> Let us consider the most influential cause for everything done by nature, the only cause that can lead to an understanding of everything that nature produces. . . . This is, in effect, a cause whose power is absolute, superior even to nature, since it regulates all nature's acts, a cause whose empire embraces all parts of nature's domain. . . . This cause resides in the power that circumstances have to modify all operations of nature, to force nature to change continually the laws that she would have followed without [the intervention of] these circumstances, and to determine the character of each of her products. The extreme diversity of nature's productions must also be attributed to this cause.

Lamarck's great intellectual journey began with a public address about evolution, delivered in 1800 during a month that the revolutionary government had auspiciously named *Floréal,* or flowering. He then developed the first comprehensive theory of evolution in modern science—an achievement that won him a secure place in any scientific hall of fame or list of immortals—despite the vicissitudes of his reputation during his own lifetime and immediately thereafter.

But Lamarck's original system failed—and not for the reasons that we usually specify today in false hindsight (the triumph of Mendelism over Lamarck's erroneous belief in inheritance of acquired characters), but by inconsistencies that new information imposed upon the central logic of Lamarck's system during his own lifetime. We can identify a fulcrum, a key moment, in the unraveling of Lamarck's original theory—when he attended a lecture by Cuvier on the anatomy of annelids, and recognized that he would have to split his taxonomic class of worms into two distinct groups. This recognition—which Lamarck recorded with excitement (and original art) as a handwritten insertion into his first published book on evolution—unleashed a growing cascade of consequences that, by Lamarck's last book of 1820, had destroyed his original theory of primary ladders of progress versus subsidiary lateral deviations, and led him to embrace the opposite model (in both geometry of animal classification, and basic philosophy of nature) of a branching tree of life.

A conventional interpretation would view this tale as fundamentally sad, if not tragic, and would surely note a remarkable symbol and irony for a literary conclusion. Lamarck began his adventure in the springtime month of flowering. But he heard Cuvier's lecture, and his system began to crumble, on the eleventh day of Nivôse—the winter month of snow. How fitting—to begin with springtime joy and promise, and to end in the cold and darkness of winter.

How fitting in one distorted sense—but how very, very wrong. I do not deny or belittle Lamarck's personal distress, but how can we view his slow acknowledgment of logical error, and his willingness to construct an entirely new and contrary explanation, as anything other than a heroic act, worthy of our greatest admiration and identifying Lamarck as one of the finest intellects in the history of biology (the name that he invented for his discipline). Two major reasons lead me to view Lamarck's intellectual odyssey in this eminently positive light. First, what can be more salutary in science than the flexibility that allows a person to change his mind—and to do so not for a minor point under the compulsion of irrefutable data, but to rethink and reverse the most fundamental concept underlying a basic philosophy of nature?

I would argue, secondly, that Lamarck's journey teaches us something vitally important about the interaction between nature and our attempts to understand her ways. The fallacies and foibles of human thinking generate systematic and predictable trouble when we try to grasp the complexities of external reality. Among these foibles, our persistent attempts to build abstractly beautiful, logically impeccable, and comprehensively simplified systems always lead us astray. Lamarck far exceeded most colleagues in his attraction to this perilous style of theorizing—this *esprit de système*—and he therefore fell further and harder because he also possessed the honesty and intellectual power to probe his mistakes.

Nature, to cite a modern cliché, always bats last. She will not succumb to the simplicities of our hopes or mental foibles, but she remains eminently comprehensible. Evolution follows the syncopated drumbeats of complex and contingent histories, shaped by the vagaries and uniquenesses of time, place, and environment. Simple laws with predictable outcomes cannot fully describe the pageant and pathways of life. A linear march of progress must fail as a model for evolution, but a luxuriantly branching tree does express the basic geometry of history.

When Lamarck snatched victory from the jaws of his defeat (by abandoning his beloved ladder of life and embracing the tree), he stood in proper humility before nature's complexity—a lesson for us all. But he also continued to wrestle with nature, to struggle to understand and even to tame her ways, not simply to bow down and acknowledge sovereignty. Only the most heroic people can follow Job's great example in owning error while continuing to hurl defiance and to shout "I am here." Lamarck greeted nature (traditionally construed as female) with Job's ultimate challenge to God (construed as male, in equally dubious tradition): "Though he slay me, yet will I trust in him; but I will maintain mine own ways before him" (Job 13:15).

I therefore propose that we reinterpret the symbolic meaning of Lamarck's undoing in the month of Nivôse. Cuvier's challenge elicited a cascade of discovery and reform, not the battering of bitter defeat. And snow also suggests metaphors of softness, whiteness, and purification—not only of frost, darkness, and destruction. God, in a much kinder mood than he showed to poor Job, promised his people in the first chapter of Isaiah: "though your sins be as scarlet, they shall be as white as snow; though they be red like crimson, they shall be as wool." We should also remember that this biblical verse begins with an even more famous statement—a watchword for an intellectual life, and a testimony to Lamarck's brilliance and flexibility: "Come now, and let us reason together."

III

Darwin's Century — and Ours

Lessons from Britain's Four Greatest Victorian Naturalists

7

Lyell's Pillars of Wisdom

I. Controlling the Fires of Vesuvius

THE TWO CLASSICAL SCENARIOS FOR A CATASTROPHIC end of all things—destruction by heat and flames or by cold and darkness—offer little fodder for extended discussion about preferences, a point embedded, with all the beauty of brevity, in Robert Frost's poem "Fire and Ice," written in 1923:

Some say the world will end in fire,
Some say in ice.
From what I've tasted of desire
I hold with those who favor fire.
But if it had to perish twice,
I think I know enough of hate
To say that for destruction ice
Is also great
And would suffice.

Among natural phenomena that poets and scholars have regarded as heralds or harbingers of the final consummation, volcanic eruptions hold pride of place. Mount Vesuvius may represent a mere pimple of activity compared with the Indonesian explosion of Tambora in 1815 or Kratakau in 1883, but a prime location on the Bay of Naples, combined with numerous eruptions at interesting times, has promoted this relatively small volcano into a primary symbol of natural terror.

Given our traditional dichotomy for unpleasant finality, I note with some amusement that the two most famous encounters of celebrated scientists with this archetypal volcano—one in each millennium of modern history—have elicited contrasting comparisons of Vesuvian eruptions with the end of time: "lights out" for the first, "up in flames" for the second.

Pliny the Elder (A.D. 23–79) wrote a massive compendium, *Natural History,* divided into thirty-seven *libri* (books) treating all aspects, both factual and folkloric, of subjects now gathered under the rubric of science. Pliny's encyclopedia exerted enormous influence upon the history of Western thought, particularly during the Renaissance (literally "rebirth"), when rediscovery of classical knowledge became the primary goal of scholarship (see chapter 3). Several editions of Pliny's great work appeared during the first few decades of printing, following the publication of Gutenberg's Bible in 1455.

In August of A.D. 79, while serving as commander of the fleet in the Bay of Naples, Pliny noted a great cloud arising from Mount Vesuvius. Following the unbeatable combination of a scientist's curiosity and a commander's duty, Pliny sailed toward the volcano, both to observe more closely and to render aid. He went ashore at a friend's villa, made a fateful decision to abandon the shaking houses for the open fields, and died by asphyxiation in the same eruption that buried the cities of Pompeii and Herculaneum.

Pliny the Younger, his nephew and adopted son, remained at their villa, a few miles farther west of the volcano, to continue (as he stated) his studies of Livy's historical texts. After the dust had settled—sorry, but I couldn't resist this opportunity to use a cliché literally—he wrote two famous letters to the historian Tacitus, describing what he had heard of his uncle's fate and what he had experienced on his own. Pliny the Younger recounted all the horrors of shaking houses, falling rocks, and noxious fumes, but he emphasized the intense darkness produced by the spreading volcanic cloud, a pall that he could compare only with one scenario for the end of time.*

*In one of those odd coincidences that make writing, and intellectual life in general, such a joy, I happened to be reading, just two days after completing this essay, a volume of Francis Bacon's complete works. I knew the old story about his death in 1626. Bacon, who loved to perform and report simple experiments of almost random import (his last and posthumous work, *Sylva sylvarum* [The forest of forests], lists exactly one thousand such observations and anecdotes), wanted to learn if

> A darkness overspread us, not like that of a cloudy night, or when there is no moon, but of a room when it is shut up and all the lights are extinguished. Nothing then was to be heard but the shrieks of women, the screams of children, and the cries of men . . . some wishing to die from the very fear of dying, some lifting up their hands to the gods; but the greater part imagining that the last and eternal night had come, which was to destroy both the gods and the world together.

Athanasius Kircher (1602–80), a German Jesuit who lived in Rome, where he served as an unofficial "chief scientist" for the Vatican, cannot be regarded as a household name today (although he served as a primary character and inspiration for Umberto Eco's novel *The Island of the Day Before*). Nonetheless, Kircher ranked among the most formidable intellects of the seventeenth century. He wrote, for example, the most famous works of his time on magnetism, music, China (where the Jesuit order had already established a major presence), and the interpretation of Egyptian hieroglyphics (his system ultimately failed, but did offer important clues and inspiration for later scholars). Kircher tumbled into intellectual limbo largely because his Neoplatonic worldview became so completely eclipsed by an alternative concept of causality that we call modern science—a reform that Galileo (whom Kircher had more or less replaced as a

---

snow could retard putrefaction. He therefore stopped his carriage on a cold winter day, bought a hen from a poultryman, and stuffed it with snow. He was then overtaken with a sudden chill that led to bronchitis. Too ill to reach London, Bacon sought refuge instead at the home of a friend, the earl of Arundel, where he died a few days later.

But I had never read Bacon's last and poignant letter, with its touching reference to Pliny the Elder's similar demise in his boots—and, in the context of this essay, the ironic likeness of icy scenarios for endings: Pliny the Younger's primary invocation of darkness, and Bacon's literal encounter with cold:

> My very good lord,
>
> I was likely to have had the fortune of Caius Plinius the elder, who lost his life by trying an experiment about the burning of the mountain Vesuvius: for I was also desirous to try an experiment or two, touching on the conversion and induration of bodies. As for the experiment itself, it succeeded excellently well; but in the journey (between London and Highgate) I was taken with such a fit of casting [an old term for vomiting, from *casting* in the sense of "throwing out or up," as in dice or a fishing line] as I know not whether it were the stone, or some surfeit [that is, kidney or gall stones, or overeating], or cold, or indeed a touch of them all three. But when I came to your lordship's house, I was not able to go back, and therefore was forced to take up my lodging here. . . . I kiss your noble hands for the welcome. . . . I know how unfit it is for me to write to your lordship with any other hand than my own, but by my troth my fingers are so disjointed with this fit of sickness, that I cannot steadily hold a pen.

leading scientist in Vatican eyes) had espoused in the generation just before, and that Newton would carry to triumph in the generation to follow.

Kircher published his masterpiece in 1664, an immense and amazing work entitled *Mundus subterraneus* (Underground world), and covering all aspects of anything that dwelled or occurred within the earth's interior—from lizards in caves, to fossils in rocks, to mountain springs, earthquakes, and volcanoes. Kircher had been inspired to write this work in 1637–38 when he witnessed the major eruptions of Etna and Stromboli. Mount Vesuvius, after centuries of quiescence, had also erupted in 1631, and Kircher eagerly awaited the opportunity to visit this most famous volcano on his return route to Rome.

He climbed the mountain at night, guided by flames still issuing from the active crater, and then lowered himself as far as he dared into the fuming and bubbling vent the next morning. When he published his great treatise twenty-five years later, the memories of his fear and wonder remained so strong that he prefaced his entire volume with a vivid personal tale of his encounter with a primary symbol for the end of time. But Kircher favored the alternative scenario of fire:

> In the middle of the night, I climbed the mount with great difficulty, moving upward along steep and rugged paths, toward the crater, which, horrible to say, I saw before me, lit entirely by fire and flowing pitch, and enveloped by noxious fumes of sulphur. . . . Oh, the immensity of divine power and God's wisdom! How incomprehensible are thy ways! If, in thy power, such fearful portents of nature now punish the duplicity and maliciousness of men, how shall it be in that last day when the earth, subjected to thy divine anger, is dissolved by heat into its elements. (My translation from Kircher's Latin)

I like to imagine that, as he wrote these lines, this greatest of priestly scientists hummed, sotto voce, the haunting Gregorian tune of the *Dies irae,* the most famous prayer about the last judgment:

> *Dies irae, dies illa*
> *Solvet saeclum in favilla*
> [On this day of anger,
> the world will dissolve into ashes]

Vesuvius looms over modern Naples even more ominously than Mount Rainier over Seattle, for Vesuvius lies much closer to the city center and sports a record of much more recent and frequent activity—though neither city could ever claim a medal from the global commission on safe geological siting. (My father, as a GI in World War II, observed the aftermath of the last eruption of Vesuvius in 1944.) In the light of this historical testimony, combined with a continuing and pervasive presence for any modern visitor (from a majestic mountain standing tall on the horizon, to the petrified bread and bathroom graffiti of ordinary life suddenly extinguished one fine day in Pompeii), how could anyone fail to draw from Vesuvius the same geological lesson that led Pliny and Kircher to extrapolate from a raging local volcano to a globally catastrophic end of time: the history of our planet must be ruled by sudden cataclysms that rupture episodes of quiescence and mark the dawn of a new order.

And yet the most famous geological invocation of volcanism in the Bay of Naples, bolstered by the most celebrated visual image in the profession's entire history, led scientific views of the earth in the opposite direction—toward a theory that currently observable processes, operating at characteristically gradual rates, could explain the full pageant of planetary history without any invocation of episodic global paroxysms or early periods of tumultuous planetary change, superseded later by staid global maturity.

Charles Lyell (1797–1875), the primary architect of this "uniformitarian" view, and the most famous name in the history of anglophonic geology, visited Naples on the "grand tour" of European cultural centers that nearly all Britons of good breeding undertook as an essential part of a gentleman's education. He made all the customary stops, from the steaming vents and bubbling pools of the Phlegraean Fields, to the early excavations of Pompeii, to the obligatory ascent of Vesuvius (still putting on a good show after erupting throughout the late eighteenth century, during the long tenure in Naples of British diplomat, and aficionado of volcanoes, Sir William Hamilton—a level of ardent activity matched only by the torrid, and rather public, affair between Hamilton's wife Emma and Lord Nelson himself).

How, then, could Lyell redefine Naples as a source of support for a theory so contrary both to traditional interpretations and to the plain meaning of the grandest local sights? This question occupied the forefront of my mind as I prepared for my first trip to Naples. In contemplating this geological mecca, I could hardly wait to visit the palpable signs of Pliny's misfortune (the excavations of Pompeii and Herculaneum), and to follow Kircher's path to their immediate source. But most of all, I wanted to stand upon the site of Lyell's visual epiphany, the source of his frontispiece for *Principles of Geology* (1830–33)—perhaps the most important scientific textbook ever written—and the primary icon for transforming the Vesuvian landscape from a poster for catastrophism into a paradoxical proof of triumphant gradualism: the three Roman columns of the so-called Temple of Serapis (actually a marketplace) at Pozzuoli. (I shall document, in the second part of this essay, how Lyell used these three pillars as a "tide gauge" to record extensive and gradual changes of land and sea levels during the past two thousand years—a uniformitarian antidote to the image of fiery Vesuvius as a symbol for catastrophic global endings.)

The clichés of travel literature require an arduous journey sparked with tales of adventure and danger. But I have never managed to strike up a friendship with this stylistic convention, and I remain a city boy at heart (and therefore quite unafraid of rather different kinds of dangers). In truth, I never got to the top of Vesuvius. My rented car carried no tire chains, and a sheet of January ice had closed the road. As for Pozzuoli, I can't claim any more adventure than a trip to South Ferry or Ozone Park would provide. Pozzuoli is the last stop on the Neapolitan subway.

But then, why should intellectual content correlate with difficulty of physical access—a common supposition that must rank among the silliest of romantic myths? Some of the greatest discoveries in the history of science have occurred in libraries or resided, unsuspected for decades, in museum drawers. By all means, take that dogsled across the frozen wastes if no alternative exists, but if the A train also goes to the same destination, why not join Duke Ellington for a smoother ride?

To reach the specifics of Pozzuoli on a literary journey, we must follow the path of Lyell's general theory. Lyell, a barrister by original profession, sought to reform the science of geology on both substantive and methodological grounds. He based his system—one might say his brief—on two fundamental propositions. *First,* the doctrine of gradualism: modern causes, operating entirely within the range of rates now observable, can explain the full spectrum of geological history. Apparently grandiose or catastrophic events truly arise by a summation of small changes through the immensity of geological time—the deep canyon carved grain by grain, the high mountain raised in numerous increments of earthquake and eruption over millions of years.

Second, the claim for a nondirectional or steady-state earth. Standard geological causes (erosion, deposition, uplift, and so on) show no trend either to increase or decrease in general intensity through time. Moreover, even the physical state of the earth (relative temperatures, positions of climatic belts, percentages of land and sea) tends to remain roughly the same, or to cycle around and around, through time. Change never slows or ceases; mountains rise and erode; seas move in and out. But the average state of the earth experiences no systematic trend in any sustained direction. Lyell even believed at first, though he changed his mind by the 1850s, when he finally concluded that mammals would not be found in the oldest strata, that the average complexity of life had remained constant. Old species die, and new species originate (by creation or by some unknown natural mechanism). But clams remain clams, and mammals mammals, from the earliest history of life until now.

When a scientist proposes such a comprehensive system, we often gain our best insights into the sources and rationale for his reforms by explicating the alternative worldview of his opponents. New theories rarely enter a previous conceptual void; rather, they arise as putative improvements or replacements for previous conventionalities. In this case, Lyell's perceived adversaries advocated an approach to geology often labeled either as catastrophism or directionalism (in opposition to Lyell's two chief tenets of gradualistic change on an earth in steady state).

Catastrophists argued that most geological change occurred in rare episodes of truly global paroxysm, marked by the "usual suspects" of volcanism, mountain building, earthquakes, flooding, and the like. Most catastrophists also held that the frequency and intensity of such episodes had decreased markedly through time, thus contrasting a feisty young earth with a much calmer planet in its current maturity.

For most catastrophists, these two essential postulates flowed logically from a single theory about the earth's history—the origin of the planet as a molten fireball spun off from the sun (according to the hypothesis, then favored, of Kant and Laplace), followed by progressive cooling. As this cooling proceeded, the outer crust solidified while the molten interior contracted continuously. The resulting instability—caused, almost literally, by an enlarging gap between the solidified crust and the contracting molten interior—eventually induced a sudden global readjustment, as the crust fractured and collapsed upon the contracted molten core. Thus, directionalism based on continuous cooling linked the catastrophism of occasional readjustment by crustal collapse with the hypothesis of a pervasive "arrow of time" leading from a fiery beginning, replete with more frequent and more intense paroxysms, to our current era of relative calm and rarer disruption.

Incidentally, this account of catastrophism as a genuine and interesting scientific alternative to Lyellian uniformity disproves the conventional canard, originally floated as a rhetorical device by Lyell and his partisans, but then incorporated uncritically as the conventional wisdom of the profession. In this Manichaean account, catastrophism represented the last stronghold for enemies of modern science: theologically tainted dogmatists who wanted to preserve both the literal time scale of Genesis and the miraculous hand of God as a prime mover by invoking a doctrine of global paroxysm to compress the grand panoply of geological change into a mere few thousand years. In fact, by the 1830s, all scientists, catastrophists and uniformitarians alike, had accepted the immensity of geological time as a central and proven fact of their emerging profession (see chapter 5). Catastrophists upheld a different theory of change on an equally ancient earth—and their views cannot be judged less "scientific," or more theologically influenced, than anything touted by Lyell and his school.

The personal, social, and scientific reasons behind Lyell's chosen commitments represent a complex and fascinating subject well beyond the scope of this essay. But we may at least note the overt strategy, chosen by this master of persuasive rhetoric, this barrister manqué, to promulgate his uniformitarian doctrine as the centerpiece of his textbook, *Principles of Geology.* In part, he chose the substantive route of arguing that the world, as revealed by geological evidence, just happens to operate by gradual and nondirectional change. But Lyell awarded primacy of place to a methodological claim: only such a uniformitarian approach, he urged, could free the emerging science of geology from previous fetters and fanciful, largely armchair, speculation.

If global paroxysms forge most of history, Lyell argued, then how can we ever develop a workable science of geology—for we have not witnessed such events in the admittedly limited duration of human history, and we can therefore identify no observational basis for empirical study. And if a tumultuous past operated so differently from a calmer present, then how can we use modern processes—the only mechanisms subject to direct observation and experiment, after all—to resolve the past? But on an earth in steady state, built entirely by modern causes acting at current intensities, the present becomes, in an old pedagogical cliché, "the key to the past," and the earth's entire history opens to scientific study. Thus, in a famous statement of advocacy, Lyell condemned catastrophism as a doctrine of despair, while labeling his uniformitarian reform as the path to scientific salvation:

> Never was there a dogma more calculated to foster indolence, and to blunt the keen edge of curiosity, than this assumption of the discordance between the former and the existing causes of change. It

> produced a state of mind unfavourable in the highest conceivable degree to the candid reception of the evidence of those minute, but incessant mutations, which every part of the earth's surface is undergoing. . . . The student, instead of being encouraged with the hope of interpreting the enigmas presented to him in the earth's structure,—instead of being prompted to undertake laborious inquiries into . . . causes now in operation, was taught to despond from the first. Geology, it was affirmed, could never rise to the rank of an exact science,—the greater number of phenomena must for ever remain inexplicable. . . .
>
> In our attempt to unravel these difficult questions, we shall adopt a different course, restricting ourselves to the known or possible operations of existing causes. . . . We shall adhere to this plan . . . because . . . history informs us that this method has always put geologists on the road that leads to truth,—suggesting views which, although imperfect at first, have been found capable of improvement, until at last adopted by universal consent. (From the introductory chapter to the third and final volume of Lyell's *Principles,* 1833)

Major intellectual struggles cannot be won by success in easy and rudimentary skirmishes. Adversaries must also be outflanked on their home ground, where superior knowledge and forces should have rendered them invincible. A new theory must meet and encompass the hardest and most apparently contradictory cases head-on. Lyell understood this principle and recognized that he would have to bring the Vesuvius of Pliny and Kircher, of Pompeii and Emma Hamilton's fire, into his uniformitarian camp—not as a prisoner, but as a proud example. No other place or subject receives even half so much attention throughout the three volumes of *Principles of Geology.*

Lyell centered his uniformitarian case for Naples and Vesuvius upon two procedural themes that embodied all his logical and literary brilliance as geology's greatest master of argument. He first invoked the cardinal geological principle of appropriate scale by pointing out that a Vesuvian eruption, while ultimately catastrophic for the baker or blacksmith of Pompeii, not only causes no planetary disruption at its own moment of maximal intensity, but then falls even further into insignificance when several hundred years of subsequent quiescence erase its memory from the populace and erode its products from the landscape.

Why, then, should such a local catastrophe serve as an unquestioned model for extrapolation to sudden global doom? Perhaps we should draw an opposite lesson from the same event: local means local—and just as the canyon deepens grain by grain, so does the mountain chain rise gradually, eruption by eruption over extended time. At most, Vesuvius teaches us that increments of gradualism may be large at human scale—the lava field versus the eroded sand grain—but still small by global standards. In 1830, Lyell summarized a long chapter, "History of the volcanic eruptions in the district around Naples," by writing:

> The vast scale and violence of the volcanic operations in Campania, in the olden time, has been a theme of declamation. . . . Instead of inferring, from analogy that . . . each cone rose in succession,—and that many years and often centuries of repose intervened between each eruption—geologists seem to have conjectured that the whole group sprung up from the ground at once, like the soldiers of Cadmus when he sowed the dragon's teeth.

Moreover, Lyell continued in closing the first volume of his tenth edition (1867), even by purely local standards, natural catastrophes usually impose only a fleeting influence upon history. Most inhabitants view Campania as a land of salubrious tranquillity. As for Vesuvius itself, even the worst natural convulsion cannot match the destructive power of human violence and venality. In a striking literary passage, Lyell reminds us that Vesuvius posed maximal danger to the Roman empire when Spartacus housed the troops of his slave revolt in the volcano's quiescent crater in 73 B.C., not when lavas and poisonous gases poured out in A.D. 79:

> Yet what was the real condition of Campania during those years of dire convulsion? "A climate," says Forsyth, "where heaven's breath smells sweet and wooingly—a vigorous and luxuriant nature unparalleled in its productions—a coast which was once the fairy-land of poets, and the favourite retreat of great men." . . . The inhabitants, indeed, have enjoyed no immunity from the calamities which are the lot of mankind; but the principal evils which they have suffered must be attributed to moral, not to physical, causes—to disastrous events over which man might have exercised a control, rather than to inevitable catastrophes which result from subterranean agency. When Spartacus encamped his army of ten thousand gladiators in

> the old extinct crater of Vesuvius, the volcano was more justly a subject of terror to Campania than it has ever been since the rekindling of its fires.

For his second theme, Lyell emphasized the importance of interpreting evidence critically, but not necessarily literally. The geological record, like most archives of human history, features more gaps than documents. (In a famous metaphor, later borrowed by Darwin for a crucial argument in *The Origin of Species,* Lyell compared the geological record to a book with very few pages preserved, of these pages few lines, of the lines few words, and of the words few letters.) Moreover, the sources of imperfection often operate in a treacherous way because information does not disappear at random, but rather in a strongly biased fashion—thus tempting us to regard some causes as dominant merely because the evidence of their action tends to be preserved, while signs of truly more important factors may differentially disappear from the record.

Lyell recognized that catastrophes usually leave their signatures, for extensive outpourings of lava, or widespread fracturing of strata by earthquakes, resist erasure from the geological record. But the publishers of time often print equally important evidence for gradual change—the few inches of sediment that may accumulate during millions of years in clear calm seas, or the steady erosion of a riverbed grain by grain—upon missing pages of the geological book. This bias not only overemphasizes the role of catastrophes in general, but may also plant the false impression that intensity of geological change has diminished through time—for if the past favors the preservation of catastrophes, while the present yields more balanced data for all modes of change, then a literal and uncritical reading of geological evidence may inspire erroneous inferences about a more tumultuous past.

Lyell summarized this crucial argument about biases of preservation in a brilliant metaphor for Mount Vesuvius. "Suppose," he writes, "we had discovered two buried cities at the foot of Vesuvius, immediately superimposed upon each other, with a great mass of tuff and lava intervening, just as Portici and Resina, if now covered with ashes, would overlie Herculaneum." (When Lyell visited the area in 1828, excavations at Herculaneum had proceeded further than those at Pompeii—hence Lyell's primary citation of a town that now ranks second to Pompeii for memorializing the destructive powers of Vesuvian eruptions.) If we read such a sequence literally, we would have to infer a history built by sudden and catastrophic changes. The remains of an Italian city, littered with modern debris of beer cans and bicycles, would overlie the strata of a

The frontispiece to Lyell's Principles of Geology, *showing the three pillars of Pozzuoli, with evidence for a substantial rise and fall of sea level in historic times.*

Roman town replete with fragments of amphoras and chariots—with only a layer of volcanic rocks between. We would then conclude that a violent catastrophe had triggered a sudden mutation from Latin to Italian, and from chariot wheels to automobile tires (for we would note the genuine relationships while missing all the intermediary stages)—simply because the evidence for nearly two thousand years of gradual transitions failed to enter a historical record strongly biased toward the preservation of catastrophic events.

A successful campaign for substantial intellectual reform also requires a new and positive symbol or icon, not just a set of arguments (as presented so far) to refute previous interpretations. Vesuvius in flames, the icon of Pliny or Kircher, must be given a counterweight—some Neapolitan image, also a consequence of Vesuvian volcanism, to illustrate the efficacy of modern causes and the extensive results produced by accumulating a series of small and gradual changes through substantial time. Lyell therefore chose the Roman pillars of Pozzuoli—an image that he used as the frontispiece for all editions of *Principles of Geology* (also as an embossed golden figure on the front cover of later editions). By assuming this status as an introductory image in the most famous geological book ever written, the pillars of Pozzuoli became icon numero uno for the earth sciences. I cannot remember ever encountering a modern textbook that does not discuss Lyell's interpretation of these three columns, invariably accompanied by a reproduction of Lyell's original figure, or by an author's snapshot from his own pilgrimage.

II. Raising (and Lowering) the Columns of Pozzuoli

In exchanging the pillars of Pozzuoli for the fires of Vesuvius as a Neapolitan symbol for the essence of geological change, Lyell made a brilliant choice and a legitimate interpretation. The three tall columns—originally interpreted as remains of a temple to Serapis (an Egyptian deity much favored by the Romans as well) but now recognized as the entranceway to a marketplace—had been buried in later sediment and excavated in 1750. The marble columns, some forty feet tall, are "smooth and uninjured to the height of about twelve feet above their pedestals." Lyell then made his key observation, clearly illustrated in his frontispiece: "Above this is a zone, about nine feet in height, where the marble has been pierced by a species of marine perforating bivalve—*Lithodomus.*"

From this simple configuration, a wealth of consequences follow—all congenial to Lyell's uniformitarian view, and all produced by the same geological agents that shaped the previously reigning icon of Vesuvius in flames. The columns, obviously, were built above sea level in the first or second century A.D. But the entire structure then became partially filled by volcanic debris, and subsequently covered by sea water to a height of twenty feet above the bases of the columns. The nine feet of marine clam holes (the same animals that, as misnamed "shipworms," burrow into piers, moorings, and hulls of vessels throughout the world) prove that the columns then stood entirely underwater to this level—for these clams cannot live above the low-tide line, and the Mediterranean Sea experiences little measurable tide in any case. The nine feet of clam borings, underlain by twelve feet of uninjured column, implies that an infill of volcanic sediments had protected the lower parts of the columns—for these clams live only in clear water.

But the bases of the columns now stand at sea level—so this twenty-foot immersion must have been reversed by a subsequent raising of land nearly to the level of original construction. Thus, in a geological moment of fewer than two thousand years, the "temple of Serapis" experienced at least two major movements of the surrounding countryside (without ever toppling the columns)—more than twenty feet down, followed by a rise of comparable magnitude. If such geological activity can mark so short a time, how could anyone deny the efficacy of modern causes to render the full panoply of geological history in the hundreds of millions of years actually available? And how could anyone argue that the earth has now become quiescent, after a more fiery youth, if the mere geological moment of historical time can witness so much mobility? Thus, Lyell presented the three pillars of Pozzuoli as a triumphant icon for both key postulates of his uniformitarian system—the efficacy of modern causes, and the relative constancy of their magnitude through time.

The notion of a geologist touring Naples, but omitting nearby Pozzuoli, makes about as much sense as a tale of a pilgrim to Mecca who visited the casbah but skipped the Kaaba. Now I admire Lyell enormously as a great thinker and writer, but I have never been a partisan of his uniformitarian views. (My very first scientific paper, published in 1965, identified a logical confusion among Lyell's various definitions of uniformity.) But my own observations of the pillars of Pozzuoli seemed only to strengthen and extend his conclusions on the extent and gradual character of geological change during historical times.

I had brought only the first edition (1830–33) of Lyell's *Principles* with me to Naples. In this original text, Lyell attributed (tentatively, to be sure) all changes in level to just two discrete and rapid events. He correlated the initial subsidence (to a level where marine clams could bore into the marble pillars) to "earthquakes which preceded the eruption of the Solfatara" (a volcanic field on the outskirts of Pozzuoli) in 1198. "The pumice and other matters ejected from that volcano might have fallen in heavy showers into the sea, and would thus immediately have covered up the lower part of the columns." Lyell then ascribed the subsequent rise of the pillars to a general swelling and uplift of land that culminated in the formation of Monte Nuovo, a volcanic mound on the outskirts of Pozzuoli, in 1538.

But at the site, I observed, with some surprise, that the evidence for changing levels of land seemed more extended and complex. I noticed the high zone of clam borings on the three columns, but evidence—not mentioned by Lyell—for another discrete episode of marine incursion struck me as even more obvious and prominent, and I wondered why I had never read anything about this event. Not only on the three major columns, but on every part of the complex (see the accompanying figure)—the minor columns at the corners of the quadrangular market area, the series of still smaller columns surrounding a circular area in the middle of the market, and even the brick walls and sides of structures surrounding the quadrangle—I noted a zone, extending two to three feet up from the marble floor of the complex and terminated by a sharp line of demarcation. Within this zone, barnacles and oyster shells remain cemented to the bricks and columns—so the distinct line on top must represent a previous high-water mark. Thus, the still higher zone of clam borings does not mark the only episode of marine incursion. This lower, but more prominent, zone of shells must signify a later depression of land. But when?

Lyell's original frontispiece (redrafted from an Italian publication of 1820), which includes the bases of the large columns, depicts no evidence for this zone. Did he just fail to see the barnacles and oysters, or did this period of marine flooding occur after 1830? I scoured some antiquarian bookstores in Naples and found several early-nineteenth-century prints of the columns (from travel books about landscapes and antiquities, not from scientific publications). None showed the lower zone of barnacles and oysters. But I did learn something interesting from these prints. None depicted the minor columns now standing both in the circular area at the center, and around the edge of the quadrangle—although these locations appear as flat areas strewn with bric-a-brac in some prints. But a later print of 1848 shows columns in the central circular area. I must therefore assume that the excavators of Pozzuoli reerected the smaller columns of the quadrangle and central circle sometime near the middle of the nineteenth century—while we know that Lyell's three major columns stood upright from their first discovery in 1749. (A fourth major column still lies in several pieces on the marble floor of the complex.)

All these facts point to a coherent conclusion. The minor columns of the central circle and quadrangle include the lower zone of barnacles and oysters. These small columns were not reerected before the mid-nineteenth century. Lyell's frontispiece, and other prints from the early nineteenth century, show the three large columns without encrusting barnacles and oysters at the base. Therefore, this later subsidence of land (or rise of sea to a few feet above modern levels) must have culminated sometime after the 1840s—thus adding further evidence for Lyell's claim of substantial and complex movements of the earth *within* the geological eye-blink of historic times.

For a few days, I thought that I had made at least a minor discovery at Pozzuoli—until I returned home (and to reality), and consulted some later editions of Lyell's *Principles,* a book that became his growing and changing child (and his lifelong source of income), reaching a twelfth edition by the time of his death. In fact, Lyell documented, in two major stages, how increasing knowledge about the pillars of Pozzuoli had enriched his uniformitarian view from his initial hypothesis of two quick and discrete changes toward a scenario of more gradual and more frequent alterations of level.

1. In the early 1830s, Charles Babbage, Lyell's colleague and one of the most interesting intellectuals of Victorian Britain (more about him later), made an extensive study of the Pozzuoli columns and concluded that both the major fall of land (to the level of the clam borings) and the subsequent rise had occurred in a complex and protracted manner through several substages, and not all at once as Lyell had originally believed. Lyell wrote in the sixth edition of 1840:

> Mr. Babbage, after carefully examining several incrustations . . . as also the distinct marks of ancient lines of water-level, visible below the zone of lithophagous perforations [holes of boring clams, in plain English], has come to the conclusion, and I think, proved, that the subsidence of the building was not sudden, or at one period only, but gradual, and by successive movements. As to the re-elevation of the depressed tract, that may also have occurred at different periods.

2. When Lyell first visited Pozzuoli in 1828, the high-water level virtually matched the marble pavement. (Most early prints, including Lyell's frontispiece, show minor puddling and flooding of the complex. Later prints, including an 1836 version from Babbage that Lyell adopted as a replacement for his original frontispiece in later editions of *Principles,* tend to depict deeper water.) In 1838, Lyell read a precise account of this modern episode of renewed subsidence—and

he then monitored this most recent change in subsequent editions of *Principles.* Niccolini, "a learned architect [who] visited the ruins frequently for the sake of making drawings," found that the complex had sunk about two feet from his first observations in 1807 until 1838, when "fish were caught every day on that part of the pavement where in 1807, there was never a drop of water in calm weather."

Lyell continued to inquire about this active subsidence—from a British colleague named Smith in 1847, from an Italian named Scacchi in 1852, and from his own observations on a last trip in 1858. Lyell acknowledged several feet of recent sinking and decided to blame the old icon of Vesuvius! The volcano had been active for nearly a hundred years, including some spectacular eruptions during Hamilton's tenure as British ambassador—after several centuries of quiescence. Lyell assumed that this current subsidence of surrounding land must represent an adjustment to the loss of so much underground material from the volcano's crater. He wrote: "Vesuvius once more became a most active vent, and has been ever since, and during the same lapse of time the area of the temple, so far as we know anything of its history, has been subsiding."

In any case, I assume that the prominent layer of encrustation by marine barnacles and oysters, unmentioned by Lyell and undepicted in all my early-nineteenth-century sources—but (to my eyes at least) the most obvious sign of former geological activity at Pozzuoli today, and far more striking, in a purely visual sense, than the higher zone of clam borings—occurred during a more recent episode of higher seas. Again, we can only vindicate Lyell's conviction about the continuing efficacy of current geological processes.

A conventional essay in the hagiographical mode would end here, with Lyell triumphant even beyond the grave and his own observations. But strict uniformity, like its old alternative of uncompromising catastrophism, cannot capture all the complexity of a rich and flexible world that says yes to at least part of most honorable extremes in human system building.

Uniformity provided an important alternative and corrective to strict catastrophism, but not the complete truth about a complex earth. Much of nature does proceed in Lyell's slow and nondirectional manner, but genuine global catastrophes have also shaped our planet's history—an idea once again in vogue, given virtual proof for triggering of the late Cretaceous mass extinction, an event that removed dinosaurs along with some 50 percent of all marine species, by the impact of an extraterrestrial body. Our city of intellectual possibilities includes many mansions, and restriction to one great house will keep us walled off from much of nature's truth.

As a closing example, therefore, let us return to Lyell's fascinating colleague, Charles Babbage (1792–1871), Lucasian professor of mathematics at Cambridge, and inventor of early calculating machines that presaged the modern digital computer. The *Encyclopaedia Britannica* ends an article on this versatile genius by writing: "He assisted in establishing the modern postal system in England and compiled the first reliable actuarial tables. He also invented a type of speedometer and the locomotive cowcatcher." So why not geology as well!

Babbage presented his studies of Pozzuoli to the Geological Society of London in 1834, but didn't publish his results until 1847 because, as he stated in a preface written in the third person, "other evocations obliged him to lay it aside"—primarily that cowcatcher, no doubt! Babbage had pursued his studies to affirm Lyell's key uniformitarian postulate, as clearly indicated in the ample subtitle of his publication: "Observations on the Temple of Serapis at Pozzuoli near Naples, with an attempt to explain the causes of the frequent elevation and depression of large portions of the earth's surface in remote periods, and to prove that those causes continue in action at the present time."

By delaying publication until 1847, Babbage needed to add an appendix to describe the recent subsidence also noted by Lyell in later editions of *Principles of Geology*. Babbage discussed the observations of Niccolini and, especially, of Smith as reported to the Geological Society of London: "Mr. Smith found the floor of the temple dry at high water in 1819, and 18 inches on it at high water in 1845." But Babbage then integrated these latest data with his previous observations on earlier changes in historical times to reach his general uniformitarian conclusions:

> The joint action of certain existing and admitted causes must necessarily produce on the earth's surface a continual but usually slow change in the relative levels of the land and water. Large tracts of its surface must be slowly subsiding through the ages, whilst other portions must be rising irregularly at various rates.

To generalize this Neapolitan conclusion, Babbage then cited the ongoing work of a young naturalist, based on entirely different phenomena from the other side of the globe: coral atolls of the tropical Pacific Ocean. This young man had not yet become the Charles Darwin whom we revere today. (Publication of *The Origin of Species* still lay twelve years in the future, and Darwin had revealed his evolutionary suspicions only to a few closest confidants, not including Babbage.) Therefore, Babbage and the scientific community of Britain knew Darwin only as a promising young naturalist who had undertaken a five-year voyage around the world, published a charming book on his adventures and three scientific volumes on the geology of South

America and the formation of coral atolls, and now labored in the midst of a comprehensive treatise, which would eventually run to four volumes, on the taxonomy of barnacles.

Darwin's theory on the origin of coral atolls surely struck his colleagues as the most important and original contribution of his early work. Darwin, labeling his explanation as the "subsidence theory" of coral reefs, explained the circular form of atolls as a consequence of subsidence of the surrounding sea floor. Reefs begin by growing around the periphery of oceanic islands. If the islands then subsided, the corals might continue to grow upward, eventually forming a ring as the central island finally disappeared below the waves.

This brilliant—and largely correct—explanation included two implications particularly favorable to Lyell and his fellow uniformitarians, hence their warm embrace for this younger colleague. First, the subsidence theory provided an excellent illustration for the efficacy and continuity of gradual change—for corals could not maintain their upward growth unless the central islands sank slowly. (Reef corals, filled with symbiotic photosynthetic algae, cannot live below the level of penetration for sunlight into oceanic waters—so any rapid subsidence would extinguish the living reefs.)

Second—and more crucial to the work of Babbage and Lyell at Pozzuoli—the large geographic range of atolls proves that major regions of the earth's crust must be subsiding, thus also implying that other regions of comparable extent must rise at the same time. Therefore, the fluctuations recorded on Pozzuoli's pillars need not represent only a local phenomenon, but may also illustrate one of the most fundamental principles of the gradualist, nondirectionalist, and uniformitarian mechanics of basic planetary behavior. In fact, and above all other implications, Darwin had emphasized his discovery that coral atolls do not form in regions with active volcanoes, while no atolls exist where volcanoes flourish in eruption. This mutual avoidance indicates that large tracts of the earth's crust, not merely local pinpoints, must be subsiding or rising in concert—with atolls as primary expressions of subsidence, and volcanoes as signs of uplift.

Babbage wrote to praise the young Darwin, but also to assert that he had reached the same uniformitarian conclusions independently, during his own studies of Pozzuoli:

> Mr. Darwin, whose voyages and travels extended from 1826 to 1836 [*sic;* the *Beagle* voyage lasted from 1831 to 1836], was gradually accumulating and arranging an immense collection of facts relating to the formation of coral and lagoon islands, as well as to the relative changes of level of land and water. In 1838 Mr. Darwin published

> his views on those subjects, from which, amongst several other very important inferences, it resulted, that he had, from a large induction of facts, arrived at exactly the same conclusion as that which it has been the chief object of this paper to account for, from the action of known and existing causes.

So far, so good—and so fair, and so just. But Babbage then proceeded further—into one of the most ludicrously overextended hypotheses ever advanced in the name of uniformitarian geology. He appended a "supplement" to his 1847 publication on the pillars of Pozzuoli entitled "Conjectures concerning the physical condition of the surface of the moon." In Babbage's day, most scientists interpreted lunar craters as volcanic cones—a catastrophic explanation that Babbage wished to challenge. He noted that a region of lunar craters would look very much like a field of earthly coral atolls standing in the bed of a vanished sea:

> The perusal of Mr. Darwin's explanation of the formation of coral reefs and of lagoon island led me to compare these islands with those conical crater-shaped mountains which cover the moon's surface; and it appears to me that no more suitable place could be found for throwing out the following conjectures, than the close of a paper in which I have endeavoured to show, that known and existing causes lead necessarily to results analogous to those which Mr. Darwin has so well observed and recorded. . . .
>
> If we imagine a sea containing a multitude of such lagoon islands to be laid dry, the appearance it would present to a spectator at the moon would strongly resemble that of a country thickly studded with volcanic mountains, having craters of various sizes. May not therefore much of the apparently volcanic aspect of the moon arise from some cause which has laid dry the bottom of a former ocean on its surface?

Babbage became bolder near the end of his commentary, as he explicitly wondered "if those craters are indeed the remains of coral lagoon islands." To be fair, Babbage recognized the highly speculative nature of his hypothesis:

> The proceeding remarks are proposed entirely as speculations, whose chief use is to show that we are not entirely without principles from which we may reason on the physical structure of the

> moon, and that the volcanic theory is not the only one by which the phenomena could be explained.

But later discoveries only underscore the irony of what may be the greatest overextension of uniformitarian preferences ever proposed by a major scientist. Babbage suggested that lunar craters might be coral atolls because he wished to confute their catastrophic interpretation as volcanic vents and mountains. Indeed, lunar craters are not volcanoes. They are formed by the even more sudden and catastrophic mechanism of meteoritic impact.

Comprehensive worldviews like uniformitarianism or catastrophism provide both joys and sorrows to their scientific supporters—the great benefits of a guide to reasoning and observation, a potential beacon through the tangled complexities and fragmentary character of nature's historical records; but also and ineluctably combined with the inevitable, ever-present danger of false assurances that can blind us to contrary phenomena right before our unseeing eyes. Lyell himself emphasized this crucial point, with his characteristic literary flair, in the closing paragraph to his discussion about the pillars of Pozzuoli—in this case, to combat the prejudice that landmasses must be rock stable, with all changes of level ascribed to movements of the sea:

> A false theory it is well known may render us blind to facts, which are opposed to our prepossessions, or may conceal from us their true import when we behold them. But it is time that the geologist should in some degree overcome those first and natural impressions which induced the poets of old to select the rock as the emblem of firmness—the sea as the image of inconstancy.

But we also know that no good deed goes unpunished and that any fine principle can turn around and bite you in the ass. Lyell had invoked this maxim about the power of false theories to emphasize that conventional preferences for catastrophism had been erroneously nurtured by the differential preservation of such evidence in our imperfect geological records. But Georges Cuvier, Lyell's French colleague, leading catastrophist, and perhaps the only contemporary who could match Lyell's literary and persuasive flair, had issued the ultimate *touché* in a central passage of the most celebrated defense for geological catastrophism—his *Discours préliminaire* of 1812.

In this manifesto, Cuvier reaches an opposite conclusion from the same valid argument about the blinding force of ordinary presuppositions. We are misled, Lyell had remarked, by the differential preservation of catastrophes in the geological record. Cuvier held, *au contraire,* that we become equally blinded by the humdrum character of daily experience. Most moments, Cuvier argues, feature no local wars or deaths, and certainly no global cataclysms. So we do not properly credit these potential forces as agents of history, even though one global paroxysm every few million years (and therefore rarely, if ever, observable in a human lifetime) can shape the pageant of life on earth. Cuvier writes:

> When the traveller voyages over fertile plains and tranquil waters that, in their courses, flow by abundant vegetation, and where the land, inhabited by many people, is dotted with flourishing villages and rich cities filled with proud monuments, he is never troubled by the ravages of war or by the oppression of powerful men. He is therefore not tempted to believe that nature has her internal wars, and that the surface of the globe has been overturned by successive revolutions and diverse catastrophes.

I must now leave these two great geological gladiators, each using the same excellent tool of reason, to battle for his own particular theory about the earth's behavior. I return then to the pillars of Pozzuoli, just down the road from the third-largest preserved amphitheater of the Roman world (where we may site those warriors for a closing image). When I visited in early January of the premillennial year of 1999, I noticed a small, modern monument at one end of the Pozzuoli complex, a chipped and neglected slab of marble festooned with graffiti scrawled over a quotation with no identifying author. But I did copy the text as a good summary, less literary to be sure than the warring flourishes of Lyell or Cuvier, but equally eloquent in support of their common principle—a good guide to any scientist, and to any person who wishes to use the greatest human gift of independent reason against the presuppositions that bind us to columns of priestly or patriotic certainty, or to mountains of cultural stolidity:

> *Cio che piu importa e che i popolo, gli uomini tutti, perdano gli istinti e le abitudini pecorili che la millenaria schiavitu ha loro ispirato ed apprendano a pensare ed agire liberamente.*
>
> [What is most important, is that the populace, all people, lose the instincts and habits of the flock, which millennia of slavery have inspired in them, and learn to think and act in freedom.]

8

A Sly Dullard Named Darwin: Recognizing the Multiple Facets of Genius

I.

MOST YOUNG MEN OF HIS TIME COULD ONLY FANTAsize; but Charles Darwin experienced the overt drama of his century's archetypal episode in a genre of personal stories that we now call "coming of age": a five-year voyage of pure adventure (and much science), circumnavigating the globe on H.M.S. *Beagle.* Returning to England at age twenty-seven, Darwin became a homebody and never again left his native land, not even to cross the English Channel. Nonetheless, his subsequent life included two internal dramas far more intense, far more portentous, and (for anyone who can move beyond the equation of swashbuckling with excitement), far more thrilling than anything he had experienced as a world traveler: first, the intellectual drama of discovering both the factuality and mechanism of evolution; and second, the emotional drama of recognizing (and relishing) the revolutionary implications of his theory of natural selection, while learning the pain that revelation would impose upon both immediate family and surrounding society.

What could possibly be more exciting than this story, set in London in 1837. The *Beagle* had docked a few months before, and Darwin now lived in town, where he courted the right contacts and worked on his specimens. He learned that his small Galápagos birds all belong to the family of finches, and not to several disparate groups, as he had thought. He never suspected this result, and had therefore not recorded the separate islands where he had collected these birds. (Theory always influences our style of collecting facts. As a creationist on the voyage itself, Darwin never imagined that the birds could have originated from a common source and then differentiated locally. According to the creationist view, all species must have been created "for" the Galápagos, and the particular island of discovery therefore held no importance. But in any evolutionary reading, and with all the birds so closely related, precise locality now mattered intensely.) He therefore tried to reconstruct the data from memory. Ironically (in view of the depth of their later enmity over evolution), he even wrote to Captain FitzRoy of the *Beagle* in order to get birds that his old boss had collected—and labeled much more carefully!

On March 14, his ornithological consultant John Gould (no relation) presented a paper at the Zoological Society, showing that the small rhea, a large flightless bird, collected by Darwin in southern Patagonia, represented a new species, not merely a geographical variant as Darwin had thought. Gould heightened Darwin's interest enormously by naming the bird *Rhea darwinii*. Janet Browne writes in her fine biography of Darwin:*

> This moment more than any other in Darwin's life . . . deserves to be called a turning point. Darwin was tantalized by the week's results. Why should two closely similar rheas agree to split the country between them? Why should different finches inhabit identical islets? The Galapagos iguana, he was further told by Thomas Bell, similarly divided themselves among the islands, and the heavily built tortoises with their individualized shells again came to mind.

---

*The original version of this essay appeared in the *New York Review of Books* as a review of Janet Browne's *Voyaging*.

Darwin now made a key analogy. (Has any truly brilliant insight ever been won by pure deduction, and not by metaphor or analogy?) Darwin realized that the different species of finches and rheas each inhabited specific territories, each adjacent to the domain of another species. If both finches and rheas replaced each other geographically, then shouldn't temporal succession also occur in continuity—that is, by evolution rather than successive creation? Darwin had collected important and entirely novel fossils of large mammals. He thought, and his expert consultant Richard Owen had affirmed, that the fossils of one creature, later named *Macrauchenia* by Owen, stood close to the modern guanaco, a modern South American mammal closely related to the llama. Darwin experienced a key flash of insight and wrote in a small private notebook: "The same kind of relation that common ostrich [rhea] bears to Petisse [the new species *Rhea darwinii*], extinct guanaco to recent; in former case position, in latter time."

Darwin had not become an evolutionist during the *Beagle* voyage, but he had fallen under the spell of gradualism and uniformity in the earth's development, a view identified with his intellectual hero, the English geologist Charles Lyell (see preceding chapter). Darwin, at this stage of his career, worked primarily as geologist, not a biologist. He wrote three books on geological subjects inspired by the *Beagle* voyage—on coral reefs, volcanic islands, and the geology of South America—but none strictly on zoology.

Lyell, well apprised of Darwin's beliefs and accomplishments, rejoiced at first in the prospect of a potential disciple, schooled in the field of nature. "How I long for the return of Darwin!" he wrote to Adam Sedgwick, Darwin's old Cambridge geology teacher. Darwin and Lyell quickly became inseparable—in part as guru and disciple, in part simply as friends. Janet Browne writes of Lyell:

> Darwin was the first naturalist to use his "Principles" effectively: Lyell's first, and in many ways his only fully committed disciple. "The idea of the Pampas going up at the rate of an inch in a century, while the western coast and Andes rise many feet and unequally, has long been a dream of mine," Lyell excitedly scrawled to him in October. "What a field you have to write on! If you cannot get here for dinner, you must if possible join the evening party."

In other words, in these crucial weeks after the return of the *Beagle,* Darwin had reached evolution by a double analogy: between geographic and temporal variation, and between geological and biological gradualism. He began to fill notebook after notebook with cascading implications. He numbered these private volumes, starting with A for more factual matters of zoology, but describing a second set, M and N, as "full of metaphysics on morals and speculations on expression." He drew a tree of life on one of the pages, and then experienced an attack of caution, writing with a linguistic touch from *Beagle* days: "Heaven knows whether this agrees with Nature—Cuidado [watch out]."

I tell this story at some length both for its intrinsic excitement, and to present an interesting tidbit that has eluded previous historians, but that professional paleontologists must recognize and relish: for those who still cherish the myth that fact alone drives any good theory, I must point out that Darwin, at his key moment of insight—making his analogy from geography to time and evolution—chose an entirely false example to illustrate his correct principle! *Macrauchenia* is not, after all, an ancestor (or even a close relative) of guanacos, but a member of a unique and extinct South American mammalian group, the Litopterna. South America was an island continent—a kind of "Superaustralia" with a fauna even richer and more bizarre than Australia's until the Isthmus of Panama rose just a few million years ago and joined the continent with North America. Several orders of large mammals, now extinct, had evolved there, including the litopterns, with lineages that converged by independent adaptation upon horses and camels of other continents.

One may not wish to become as cavalier as Charles's brother Erasmus, who considered *The Origin of Species* "the most interesting book" he had ever read, and who wrote of any factual discrepancy: "The *a priori* reasoning is so entirely satisfactory to me that if the facts won't fit in, why so much for the facts is my feeling." Still, beautiful (and powerful) theories can rarely be killed by "a nasty, ugly little fact" of T. H. Huxley's famous statement—nor should major ideas be so destroyed in a recalcitrant world where reported "facts" so often turn out to be wrong. Fact and theory interact in wondrously complex, and often mutually reinforcing, ways. Theories unsupported by fact may be empty (and, if unsupportable in principle, meaningless in science): but we cannot even know where to look without some theory to test. As Darwin wrote in my favorite quotation: "How can anyone not see that all observation must be for or against some view if it is to be of any service." Whatever the historical interest in this tale, and despite the irony of the situation, we do not denigrate Darwin's achievement, or evolution's truthful power, by noting that Darwin's crucial analogy, at his moment of eureka, rested upon a factual error.

This issue of interaction between fact and theory brings us to the core of fascination with Darwin's biography. Darwin worked as an accumulator of facts nonpareil—in part because he had found the right theory and therefore knew where to look; in part as a consequence of his obsessive thoroughness; in part as a benefit of his personal wealth and connections. But he also developed one of the most powerful and integrative theoretical constructions—and surely the most disturbing to traditional views about the meaning of human life—in Western history: natural selection. How could Darwin accomplish so much? He seems so unlikely a candidate.

II.

In addition to general benefits conferred by wealth and access to influential circles, Darwin enjoyed specific predisposing advantages for becoming the midwife of evolution. His grandfather Erasmus had been a famous writer, physician, and freethinker. (In the first sentence of his preface to Mary Shelley's *Frankenstein,* P. B. Shelley had, in order to justify Dr. Frankenstein's experiment, alluded to Erasmus Darwin's atheistical view on the possibility of quickening matter by electricity.) Erasmus died before Charles's birth, but the grandson studiously read and greatly admired his grandfather's writing—and Erasmus Darwin had been a thoroughgoing evolutionist. Charles studied medicine in Edinburgh, where he became close to his teacher Robert Grant, a committed Lamarckian evolutionist delighted to have Erasmus's grandson as a student. And then, of course, Darwin enjoyed the grandest privilege of five years' exposure to nature's diversity aboard the *Beagle.* Still, he remained a creationist, if suffused with nascent doubt, when he returned to London in 1836.

Some people display their brilliance in their cradles—as with Mill learning classics and Mozart writing symphonies almost before either could walk. We are not surprised when such men become "geniuses"; in fact, we expect such an eventual status, unless illness or idiosyncrasy conquers innate promise. But descriptions of Darwin's early years could lead only to a prediction of a worthy, but undistinguished life. Absolutely nothing in any record documents the usual accoutrements of intellectual brilliance. Geniality and fecklessness emerge as Darwin's most visible and distinctive traits. "He was so quiet," Janet Browne writes, "that relatives found it difficult to say anything about his character beyond an appreciative nod towards an exceedingly placid temperament. Geniality was what was most often remembered by Darwin's schoolfriends: the good-humored acquiescence of an inward-looking boy who did not appear much to mind whatever happened in life. . . . Some could barely remember Darwin when asked for anecdotes at the close of his life."

Darwin did develop a passion for natural history, expressed most keenly in his beetle collection—but so many children, then and now, become total devotees to such a hobby for a transient moment in a life leading elsewhere. No one could have predicted *The Origin of Species* from a childhood insect collection. Darwin performed as an indifferent student in every phase of his formal education. Sickened by the sight of blood, he abandoned medical studies in Edinburgh. His father became so frustrated when Charles quit Edinburgh that he admonished his son: "You care for nothing but shooting, dogs, and rat-catching, and you will be a disgrace to yourself and all your family." Charles recounted the episode in his *Autobiography,* written late in life with characteristic Victorian distance and emotional restraint: "He was very properly vehement against my turning an idle sporting man, which then seemed my probable destination."

Robert Waring Darwin therefore sent his unpromising boy to Cambridge, where he could follow the usual course for unambitious later-born sons and train for the sinecure of a local parsonage. Charles showed the same interest in religion that he manifested at the time for all other academic subjects save natural history—none at all. He went along, *faute de mieux,* in his usual genial and feckless way. He obtained the Victorian equivalent of a "gentleman's C" degree, spending most of his time gambling, drinking, and hunting with his upper-class pals. He still planned to become a minister during the entire *Beagle* voyage—though I am quite sure that his thoughts always focused upon the possibilities for amateur work in natural history that such a job provided, and not at all upon the salvation of souls, or even the weekly sermon.

The *Beagle* worked its alchemy in many ways, mostly perhaps in the simple ontogenetic fact that five years represents a lot of living during one's mid-twenties and tends to mark a passage to maturity. Robert Waring Darwin, apprised by scientific colleagues of his son's remarkable collections and insights, surrendered to the inevitable change from religion to science. Charles's sister Susan wrote as the *Beagle* sailed home: "Papa and we often cogitate over what you will do when you return, as I fear there are but small hopes of your still going in the church—I think you must turn professor at Cambridge."

But the mystery remains. Why Darwin? No one thought him dull, but no one marked him as brilliant either. And no one discerned in him that primary emotional correlate of greatness that our modern age calls "fire in the belly." Thomas Carlyle, a good judge, who knew both Darwin brothers Charles and Erasmus well, considered Erasmus as far superior in intelligence.

I believe that any solution to this key puzzle in Darwinian biography must begin with a proper exegesis of intelligence—one that rejects Charles Spearman's old notion of a single scalar quantity recording overall mental might (called *g* or general intelligence, and recently revived by Murray and Herrnstein as the central fallacy of their specious book, *The Bell Curve*—see the second edition of my book *The Mismeasure of Man*). Instead, we need a concept of intelligence defined as a substantial set of largely independent attributes. This primary alternative to *g* has its own long and complex history, from an extreme in misuse by the old phrenologists, to modern tenable versions initiated by Louis L. Thurstone and J. P. Guilford, and best represented today by the work of Howard Gardner.

I do not know what *g*-score might be given to Darwin by a modern Spearmanian. I do know, however, that we must follow the alternative view of independent multiplicity to grasp Darwin's triumph in the light of such unpromising beginnings (unpromising in the apparently hopeless sense of little talent with maximal opportunity, rather than the more tractable Horatio Alger mode of great promise in difficult circumstances).

Moreover, the theory of multiplicity has an important historical and philosophical consequence for understanding human achievement. If Spearman had been correct, and if intelligence could be construed as a single, innately provided, and largely invariant scalar quantity that could be plotted as a single linear order, then we might frame a predictive and largely biological model of achievement with a predominant basis in bloodlines and a substrate in neurology. But the theory of multiplicity demands an entirely different style of biographical analysis that builds a totality from narrative details.

If the sum of a person's achievement must be sought in a subtle combination of differing attributes, each affected in marvelously varying ways by complexities of external circumstances and the interplay of psyche and society, then no account of particular accomplishment can be drawn simply by prediction based on an overall inherited mental rank. Achieved brilliance must arise from (1) a happy combination of fortunate strength in several independent attributes, joined with (2) an equally fortuitous combination of external circumstances (personal, familial, societal, and historical), so that (3) such a unique mental convergence can lead one mind and personality to solve a major puzzle about the construction of natural reality. Explanations of this kind can only be achieved in the mode of dense narrative. No shortcuts exist; the answer lies in a particular concatenation of details—and these must be elucidated and integrated descriptively.

I used to think that the last section of Darwin's autobiography (on "mental qualities") represented little more than a lie, enforced by conventions of Victorian public modesty, since Darwin could not speak openly about his strengths. The very last line may indeed be regarded as a tad disingenuous:

"With such moderate abilities as I possess, it is truly surprising that I should have influenced to a considerable extent the belief of scientific men on some important points."

In rereading this section while writing this essay, I have changed my mind. I now view Darwin's assessment of his strengths and weaknesses as probably quite accurate, but set in the false context of his own belief in something close to a Spearmanian definition of brilliance. He had internalized a fairly stereotypical notion of an acme in scientific reasoning (based largely upon mathematical ability and lightning-quick powers of deduction), and he recognized that he possessed no great strength in these areas. He understood what he could do well, but granted these powers only secondary rank and importance. If Darwin had embraced the notion of intelligence as a plethora of largely independent attributes, and had also recognized (as he did for the evolution of organisms) that great achievement also requires a happy concatenation of uncontrollable external circumstances, then he might not have been so surprised by his own success.

Darwin begins the last section of his autobiography with deep regret for his negatives:

> I have no great quickness of apprehension or wit which is so remarkable in some clever men, for instance, Huxley. . . . My power to follow a long and purely abstract train of thought is very limited; and therefore I could never have succeeded with metaphysics or mathematics.

He then almost apologizes for his much humbler positive qualities:

> Some of my critics have said, "Oh, he is a good observer, but he has no power of reasoning!" I do not think that this can be true, for the 'Origin of Species' is one long argument from the beginning to the end, and it has convinced not a few able men. No one could have written it without having some power of reasoning. I think that I am superior to the common run of men in noticing things which easily escape attention, and in observing them carefully. My industry has been nearly as great as it could have been in the observation and collection of facts. What is more important, my love of natural science has been steady and ardent. . . . From my early youth I have had the strongest desire to understand or explain whatever I

> observed. . . . These causes combined have given me the patience to reflect or ponder for any number of years over any unexplained problem.

The beginning of the final paragraph beautifully summarizes the argument that I advocate here—sublime achievement as a unique joining, at a favorable social and historical moment, of a synergistic set of disparate mental attributes. But Darwin does not accept this definition and therefore views his achievement—which he does not deny, for he was not, internally at least, a modest man—as something of a puzzle:

> Therefore my success as a man of science, whatever this may have amounted to, has been determined, as far as I can judge, by complex and diversified mental qualities and conditions.

Janet Browne did not write her book with such a theory of intelligence and accomplishment explicitly in mind, but her biography of Darwin explains his achievements better than any previous work because she has provided the requisite thick description of both the various attributes of mind (multiple intelligences) that motivated Darwin's work and powered his conclusions, and the conjunction of numerous external factors that fed his triumph.

Darwin's multiple intelligences: As the greatest veil-coverers of recent times, the Victorians did not only hide their sexual habits. More generally, they concealed most displays of passion about anything. Since passion may be the common ground for Darwin's diverse strengths, and since he so carefully constructed an external persona of dispassionate gentility, this wellspring of his greatness can easily be missed. But the sheer accumulative density of Browne's documentation eventually prevails.

We come to understand, first of all, Darwin's enormous energy—whether overt and physical during his active youth on the *Beagle,* or cerebral when he became an invalid for most of his adult life. (Some people just seem to live at a higher plane of intensity, and must see most of us as we view the languorous world of a sloth—see chapter 4 on Buffon.) We often miss this theme with Darwin because he led such a quiet life as an adult and spent so much time prostrated by illness. But I am, of course, speaking about an internal drive. Our minds are blank or unproductive most of the time (filled with so much Joycean buzz that we can't sort out a useful theme). Darwin must have been thinking with useful focus all the time, even on his sickbed. I don't quite understand how this intense energy meshes with Darwin's placidity of personality (as expressed so strongly from earliest days), the geniality that makes him so immensely likable among history's geniuses—usually a far more prickly lot. Perhaps he just kept the prickly parts under wrap because he had been schooled as such an eminent Victorian. Perhaps (a more likely alternative, I think) emotional placidity and level of intrinsic energy just represent different and separable aspects of human personalities.

In any case, this "energy"—expressed as passion, range, thoroughness, zeal, even ruthlessness at times—drove Darwin's achievements. He expressed the most overt form in roving over South America, trekking for weeks across mountains and deserts because he heard some rumor about fossil bones at the other end—and then, with equal restlessness, thinking and thinking about his results until he could encompass them in a broad theoretical conception. (For example, Darwin developed a correct theory for the origin of coral reefs—his first great contribution to science—by reading and pondering before he ever reached Pacific atolls for direct observation.)

Back in London, Darwin virtually moved into the Athenaeum Club by day, using its excellent library as his private preserve, reading the best books on all subjects from cover to cover. Browne writes:

> Something of Darwin's mettle also showed through in the way he set off on a lifetime's program of reading in areas formerly holding only faint attractions. He tackled David Hume, Adam Smith, and John Locke in turn; Herbert Mayo's *Philosophy of Living* (1837), Sir Thomas Browne's *Religio Medici,* and John Abercrombie's *Inquiries Concerning the Intellectual Powers and the Investigation of Truth* (1838) came between Gibbon and Sir Walter Scott.

As Darwin read theory and philosophy in all fields, he also began an almost obsessive querying and recording of anyone in any station who might supply information about natural history:

> He asked Mark, Dr. Darwin's [his father's] coachman, for his opinion on dogs, and Thomas Eyton for his views on owls and pigs. He made Fox [his cousin] struggle with a deluge of farmyard questions of all shapes and sizes. He struck up a correspondence with his Uncle Jos about Staffordshire worms. . . . Darwin elaborated this way of proceeding into one of the most distinctive aspects of his

> life's work. When seeking information on any new topic, he learned to go straight to the breeders and gardeners, the zookeepers, Highland ghillies, and pigeon fanciers of Victorian Britain, who possessed great practical expertise and, as Darwin fondly imagined, hardly any interest in pursuing larger theoretical explanations. . . . Being a gentleman—being able to use his social position to draw out material from people rarely considered scientific authorities in their own right—was important. His notebooks began bulging with details methodically appropriated from a world of expertise normally kept separate from high science.

Darwin pushed himself through these intense cycles of reading, pondering, noting, asking, corresponding, and experimenting over and over again in his career. He proceeded in this way when he wrote four volumes on the taxonomy of barnacles in the late 1840s and mid-1850s, when he experimented on the biogeography of floating seeds in the 1850s, on fertilization of orchids by insects in the early 1860s, and when he bred pigeons, studied insectivorous and climbing plants, and measured rates of formation of soil by worms.

He coordinated all the multiple intelligences that can seek, obtain, and order such information with his great weapon, a secret to all but a few within his innermost circle until he published *The Origin of Species* in 1859—the truth of evolution explained by the mechanism of natural selection. I don't think that he could have made sense of so much, or been able to keep going with such concentration and intensity, without a master key of this kind. He must have used yet another extraordinary intelligence to wrest this great truth from nature. But once he did so, he could then bring all his other mentalities to bear upon a quest never matched for expansiveness and import: to reformulate all understanding of nature, from bacterial physiology to human psychology, as a history of physical continuity, "descent with modification" in his words. Had he not noted, with justified youthful hubris in an early notebook, "He who understands baboon would do more towards metaphysics than Locke"?

Darwin's fortunate circumstances: All the world's brilliance, and all the soul's energy, cannot combine to produce historical impact without a happy coincidence of external factors that cannot be fully controlled: health and peace to live into adulthood, sufficient social acceptability to gain a hearing, and life in a century able to understand (though not necessarily, at least at first, to believe). George Eliot, in the preface to *Middlemarch,* wrote about the pain of brilliant women without opportunity:

> Here and there a cygnet is reared uneasily among the ducklings in the brown pond, and never finds the living stream in fellowship with its own oary-footed kind. Here and there is born a Saint Theresa, foundress of nothing, whose loving heartbeats and sobs after an unattained goodness tremble off and are dispersed among hindrances instead of centering in some long-recognizable deed.

Darwin experienced the good fortune of membership in that currently politically incorrect but ever-so-blessed group—upper-class white males of substantial wealth and maximal opportunity. This subject, a staple of recent Darwinian biographies, has been particularly well treated by Adrian Desmond and James Moore in their biography, *Darwin* (Warner Books, 1992). Though the theme is now familiar to me, I never cease to be amazed by the pervasive, silent, and apparently frictionless functioning (in smoothness of operation, not lack of emotionality) of the Victorian gentleman's world—the clubs, the networks, the mutual favors, the exclusions of some people, with never a word mentioned. Darwin just slid into this world and stuck there. He used his wealth, his illnesses, his country residence, his protective wife for one overarching purpose: to shield himself from ordinary responsibility and to acquire precious time for intellectual work. Darwin knew what he was doing and wrote in his autobiography: "I have had ample leisure from not having to earn my own bread. Even ill-health, though it has annihilated several years of my life, has saved me from the distractions of society and amusement."

Janet Browne's greatest contribution lies in her new emphasis upon a theme that has always been recognized but, strangely, never exploited as a major focus for a Darwin biography—the dynamics of immediate family. I had never realized, for example, just how wealthy and powerful Darwin's father had been. I had known that he was a famous physician, but I hadn't appreciated his role as the most prominent moneylender in the county. He was a fair and patient man, but nearly everyone who was anyone owed him something. And I feel even more enlightened about the warm and enabling (in the good sense) relationship of Charles and his remarkable wife, Emma, whose unwritten biography remains, in my view, one of the strangest absences in our scholarship about nineteenth-century life. (Sources exist in abundance for more than a few Ph.D.s. We even know the cumulative results of thirty years of nightly backgammon games between Charles and Emma; for Charles, as I have said, was an ardent recorder of details!) Much has been written about Emma and other family ties as sources of Charles's hesitancies and cautions (and I accept, as substantially true, the old cliché that Charles delayed publication because he feared the impact of his freethinking ideas upon the psyche of his devout wife). But we need to give more attention and study to Darwin's family as promoters of his astonishing achievement.

If I had to summarize the paradoxes of Darwin's complex persona in a phrase, I would say that he was a philosophical and scientific radical, a political liberal, and a social conservative (in the sense of lifestyle, not of belief)—and that he was equally passionate about all three contradictory tendencies. Many biographers have argued that the intellectual radical must be construed as the "real" Darwin, with the social conservative as a superficial aspect of character, serving to hide an inner self and intent. To me, this heroically Platonic view can only be labeled as nonsense. If a serial killer has love in his heart, is he not a murderer nonetheless? And if a man with evil thoughts works consistently for the good of his fellows, do we not properly honor his overt deeds? All the Darwins build parts of a complex whole; all are equally him. We must acknowledge all facets to fully understand a person, and not try to peel away layers toward a nonexistent archetypal core. Darwin hid many of his selves consummately well, and we shall have to excavate if we wish to comprehend. I, for one, am not fazed. Paleontologists know about digging.

9

An Awful Terrible Dinosaurian Irony

STRONG AND SUBLIME WORDS OFTEN LOSE THEIR sharp meanings by slipping into slangy cuteness or insipidity. Julia Ward Howe may not win History's accolades as a great poet, but the stirring first verse of her "Battle Hymn" will always symbolize both the pain and might of America's crucial hour:

Mine eyes have seen the glory of the coming of the Lord;
He is trampling out the vintage where the grapes of wrath are stored;
He hath loosed the fateful lightning of his terrible, swift sword;
His truth is marching on.

The second line, borrowed from Isaiah 63:3, provided John Steinbeck with a title for the major literary marker of another troubled time in American history. But the third line packs no punch today, because *terrible* now means "sorta scary" or "kinda sad"—as in "Gee, it's terrible your team lost today." But *terrible,* to Ms. Howe and her more serious age, embodied the very opposite of merely mild lament. *Terrible*—one of the harshest words available to Victorian writers—invoked the highest form of fear, or "terror," and still maintains a primary definition in *Webster's* as "exciting extreme alarm," or "overwhelmingly tragic."

Rudyard Kipling probably was a great poet, but "Recessional" may disappear from the educational canon for its smug assumption of British superiority:

God of our fathers, known of old,
Lord of our far-flung battle line,
Beneath whose awful Hand we hold
Dominion over palm and pine.

For Kipling, "awful Hand" evokes a powerful image of fearsome greatness, an assertion of majesty that can only inspire awe, or stunned wonder. Today, an awful hand is only unwelcome, as in "keep your awful hand off me"—or impoverishing, as in your pair of aces versus his three queens.

Unfortunately, the most famous of all fossils also suffer from such a demotion of meaning. Just about every aficionado knows that *dinosaur* means "terrible lizard"—a name first applied to these prototypes of prehistoric power by the great British anatomist Richard Owen in 1842. In our culture, reptiles serve as a prime symbol of slimy evil, and scaly, duplicitous, beady-eyed disgust—from the serpent that tempted Eve in the Garden, to the dragons killed by Saint George or Siegfried. Therefore, we assume that Owen combined the Greek *deinos* (terrible) with *sauros* (lizard) to express the presumed nastiness and ugliness of such a reprehensible form scaled up to such huge dimensions. The current debasement of *terrible* from "truly fearsome" to "sorta yucky" only adds to the negative image already implied by Owen's original name.

In fact, Owen coined his famous moniker for a precisely opposite reason. He wished to emphasize the awesome and fearful majesty of such astonishingly large, yet so intricate and well-adapted creatures, living so long ago. He therefore chose a word that would evoke maximal awe and respect—*terrible,* used in exactly the same sense as Julia Ward Howe's "terrible swift sword" of the Lord's martial glory. (I am, by the way, not drawing an inference in making this unconventional claim, but merely reporting what Owen actually said in his etymological definition of dinosaurs.)*

Owen (1804–92), then a professor at the Royal College of Surgeons and at the Royal Institution, and later the founding director of the newly independent

*A longstanding scholarly muddle surrounding the where and when of Owen's christening has been admirably resolved in two recent articles by my colleague Hugh Torrens, a geologist and historian of science at the University of Keele in England—"Where did the dinosaur get its name?" *New*

natural history division of the British Museum, had already achieved high status as England's best comparative anatomist. (He had, for example, named and described the fossil mammals collected by Darwin on the *Beagle* voyage.) Owen was a complex and mercurial figure—beloved for his wit and charm by the power brokers, but despised for alleged hypocrisy, and unbounded capacity for ingratiation, by a rising generation of young naturalists, who threw their support behind Darwin, and then virtually read Owen out of history when they gained power themselves. A recent biography by Nicolaas A. Rupke, *Richard Owen: Victorian Naturalist* (Yale University Press, 1994), has redressed the balance and restored Owen's rightful place as brilliantly skilled (in both anatomy and diplomacy), if not always at the forefront of intellectual innovation.

Owen had been commissioned, and paid a substantial sum, by the British Association for the Advancement of Science to prepare and publish a report on British fossil reptiles. (The association had, with favorable outcome, previously engaged the Swiss scientist Louis Agassiz for an account of fossil fishes. They apparently took special pleasure in finding a native son with sufficient skills to tackle these "higher" creatures.) Owen published the first volume of his reptile report in 1839. In the summer of 1841, he then presented a verbal account of his second volume at the association's annual meeting in Plymouth. Owen published the report in April 1842, with an official christening of the term *Dinosauria* on page 103:

> The combination of such characters . . . all manifested by creatures far surpassing in size the largest of existing reptiles will, it is presumed, be deemed sufficient ground for establishing a distinct tribe or suborder of Saurian Reptiles, for which I would propose the name of Dinosauria.

(From Richard Owen, *Report on British Fossil Reptiles, Part II,* London, Richard and John E. Taylor, published as *Report of the British Association for the Advancement of Science for 1841,* pages 60–204.)

Many historians have assumed that Owen coined the name in his oral presentation of 1841, and have cited this date as the origin of dinosaurs. But as Torrens shows, extensive press coverage of Owen's speech proves that he then included all dinosaur genera in his overall discussion of lizards, and had not yet chosen either to separate them as a special group, or to award them a definite

Scientist, volume 134, April 4, 1992, pages 40–44; and "Politics and paleontology: Richard Owen and the invention of dinosaurs," in J. O. Farlow and M. K. Brett-Surman, editors, *The Complete Dinosaur,* Indiana University Press, 1997, pages 175–90.

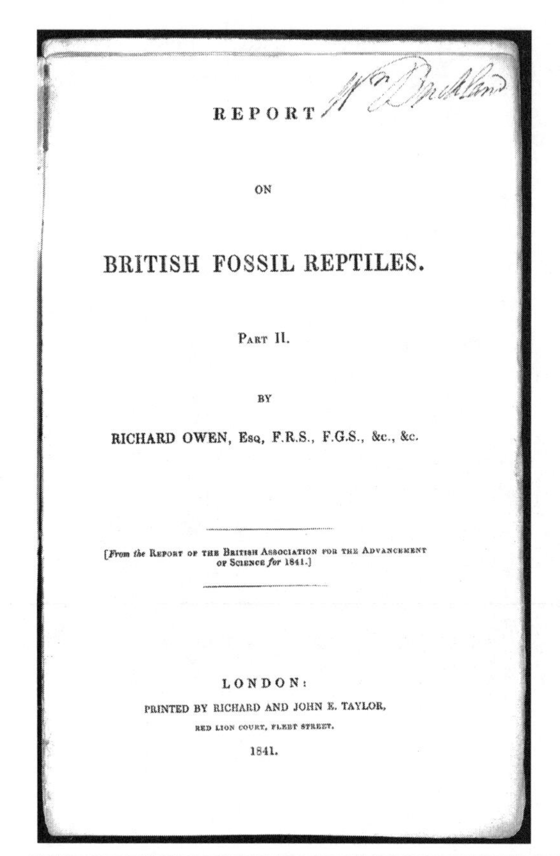

REPORT

ON

BRITISH FOSSIL REPTILES.

PART II.

BY

RICHARD OWEN, Esq., F.R.S., F.G.S., &c., &c.

[*From the* REPORT OF THE BRITISH ASSOCIATION FOR THE ADVANCEMENT OF SCIENCE *for* 1841.]

LONDON:

PRINTED BY RICHARD AND JOHN E. TAYLOR,

RED LION COURT, FLEET STREET.

1841.

William Buckland's personal copy of Richard Owen's initial report on dinosaurs. Note Buckland's signature on the title page (left) and his note (right) showing his concern with the subject of evolution ("transmutation").

name. ("Golden age" myths are usually false, but how I yearn for a time when *local* newspapers—and Torrens got his evidence from the equivalent of the *Plymouth Gazette* and the *Penzance Peeper,* not from major London journals—reported scientific talks in sufficient detail to resolve such historical questions!) Owen therefore must have coined his famous name as he prepared the report for printing—and the resulting publication of April 1842 marks the first public appearance of the term *dinosaur.* As an additional problem, a small initial run of the publication (printed for Owen's own use and distribution) bears the incorrect date of 1841 (perhaps in confusion with the time of the meeting itself, perhaps to "backdate" the name against any future debate about priority, perhaps just as a plain old mistake with no nefarious intent)—thus confounding matters even further.

In any case, Owen appended an etymological footnote to his defining words cited just above—the proof that he intended the *dino* in *dinosaur* as a mark of awe and respect, not of derision, fear, or negativity. Owen wrote: "Gr. [Greek] *deinos* [Owen's text uses Greek letters here], fearfully great; *sauros,* a lizard." Dinosaurs, in other words, are awesomely large ("fearfully great"), thus inspiring our admiration and respect, not terrible in any sense of disgust or rejection.

I do love the minutiae of natural history, but I am not so self-indulgent that I would impose an entire essay upon readers just to clear up a little historical matter about etymological intent—even for the most celebrated of all prehistoric critters. On the contrary: a deep and important story lies behind Owen's conscious and explicit decision to describe his new group with a maximally positive name marking their glory and excellence—a story, moreover, that cannot be grasped under the conventional view that dinosaurs owe their name to supposedly negative attributes.

Owen chose his strongly positive label for an excellent reason—one that could not possibly rank as more ironic today, given our current invocation of dinosaurs as a primary example of the wondrous change and variety that evolution has imparted to the history of life on our planet. In short, Owen selected his positive name in order to use dinosaurs as a focal argument *against* the most popular version of evolutionary theory in the 1840s. Owen's refutation of evolution—and his invocation of newly minted dinosaurs as a primary example—forms the climax and central point in his concluding section of a two-volume report on British fossil reptiles (entitled "Summary," and occupying pages 191–204 of the 1842 publication).

This ironic tale about the origin of dinosaurs as a weapon against evolution holds sufficient interest for its own immediate cast of characters (involving, in equal measure, the most important scientists of the day, and the fossils deemed most fascinating by posterity). But the story gains even more significance by illustrating a key principle in the history of science. All major discoveries suffer from simplistic "creation myths" or "eureka stories"—that is, tales about momentary flashes of brilliantly blinding insight by great thinkers. Such stories fuel one of the primal legends of our culture—the lonely persecuted hero, armed with a sword of truth and eventually prevailing against seemingly insuperable odds. These sagas presumably originate (and stubbornly persist against contrary evidence) because we so strongly want them to be true.

Well, sudden conversions and scales falling from eyes may work for religious epiphanies, as in the defining tale about Saul of Tarsus (subsequently renamed Paul the Apostle) on the Damascus Road: "And as he journeyed, he came near Damascus: and suddenly there shined round about him a light from heaven: And he fell to the earth, and heard a voice saying unto him, Saul, Saul, why persecutest thou me? . . . And immediately there fell from his eyes as it had been scales; and he received sight forthwith, and arose, and was baptized"* (Acts 9:4). But scientific discoveries are deep, difficult, and complex. They require a rejection of one view of reality (never an easy task, either conceptually or psychologically), in favor of a radically new order, teeming with consequences for everything held precious. One doesn't discard the comfort and foundation of a lifetime so lightly or suddenly. Moreover, even if one thinker experiences an emotional and transforming eureka, he must still work out an elaborate argument, and gather extensive empirical support, to persuade a community of colleagues often stubbornly committed to opposite views. Science, after all, operates both as a social enterprise and an intellectual adventure.

A prominent eureka myth holds that Charles Darwin invented evolution within the lonely genius of his own mind, abetted by personal observations made while he lived on a tiny ship circumnavigating the globe. He then, as the legend continues, dropped the concept like a bombshell on a stunned and shocked world in 1859. Darwin remains my personal hero, and *The Origin of Species* will always be my favorite book—but Darwin didn't invent evolution and would never have persuaded an entire intellectual community without substantial priming from generations of earlier evolutionists (including his own grandfather). These forebears prepared the ground, but never devised a plausible mechanism (as Darwin achieved with the principle of natural selection), and they never recorded, or even knew how to recognize, enough supporting documentation.

We can make a general case against such eureka myths as Darwin's epiphany, but such statements carry no credibility without historical counterexamples. If we can show that evolution inspired substantial debate among biologists long before Darwin's publication, then we obtain a primary case for interesting and extended complexity in the anatomy of an intellectual revolution. Historians have developed many such examples (and pre-Darwinian evolutionism has long been a popular subject among scholars), but the eureka myth persists, perhaps because we so yearn to place a name and a date upon defining episodes in our history. I know of no better example, however little known and poorly documented, than Owen's invention of the name *dinosaur* as an explicit weapon—ironically for the wrong side, in our current and irrelevant judgment—in an intense and public debate about the status of evolution.

*Since this essay focuses on the changing meaning of words, I just can't resist citing the next line (Acts 9:5) after the ellipsis in my quotation—surely the unintentionally funniest biblical verse based on the passage of a word from high culture into slang between the seventeenth-century King James Bible and our current vernacular: "And he said, Who art thou, Lord? And the Lord said, I am Jesus, whom thou persecutest: it is hard for thee to kick against the pricks" (then meaning "obstacles," as symbolized by sharp thorns on bushes).

Scattered observations of dinosaur bones, usually misinterpreted as human giants, pervade the earlier history of paleontology, but the first recognition of giant terrestrial reptiles from a distant age before mammalian dominance (the marine ichthyosaurs and plesiosaurs had been defined a few years earlier) did not long predate Owen's christening. In 1824, the Reverend William Buckland, an Anglican divine by title, but a leading geologist by weight of daily practice and expertise, named the first genus that Owen would eventually incorporate as a dinosaur—the carnivorous *Megalosaurus*.

Buckland devoted his professional life to promoting paleontology and religion with equal zeal. He became the first officially appointed geologist at Oxford University, and presented his inaugural lecture in 1819 under the title "Vindiciae geologicae; or the connexion of geology with religion explained." Later, in 1836, he wrote one of the eight Bridgewater Treatises, a series generously endowed by the earl of Bridgewater upon his death in 1829, and dedicated to proving "the power, wisdom and goodness of God as manifested in the creation." Darwin's circle referred to this series as the "bilgewater" treatises, and the books did represent a last serious gasp for the venerable, but fading, doctrine of "natural theology" based on the so-called "argument from design"—the proposition that God's existence, and his attributes of benevolence and perfection, could both be inferred from the good design of material objects and the harmonious interaction among nature's parts (read, in biological terms, as the excellent adaptations of organisms, and the harmony of ecosystems expressed as a "balance of nature").

Buckland (1784–1856) provided crucial patronage for several key episodes in Owen's advance, and Owen, as a consummate diplomat and astute academic politician, certainly knew and honored the sources of his favors. When Owen named dinosaurs in 1842, the theoretical views of the two men could only evoke one's favorite metaphor for indistinction, from peas in a pod, to Tweedledum and Tweedledee. Owen later became an evolutionist, though never a supporter of Darwinian natural selection. One might be cynical, and correlate Owen's philosophical shift with the death of Buckland and other powerful men of the old guard, but Owen was too intelligent (and at least sufficiently honorable) to permit such a simple interpretation, and his later evolutionary views show considerable subtlety and originality. Nonetheless—in a point crucial to this essay—Owen remained an unreconstructed Bucklandian creationist, committed to the functionalist approach of the argument from design, when he christened dinosaurs in 1842.

Gideon Mantell, a British surgeon from Sussex, and one of Europe's most skilled and powerful amateur naturalists, named a second genus (that Owen would later include among the dinosaurs) in 1825—the herbivorous *Iguanodon,* now classified as a duckbill. Later, in 1833, Mantell also named *Hylaeosaurus,* now viewed as an armored herbivorous dinosaur ranked among the ankylosaurs.

Owen united these three genera to initiate his order Dinosauria in 1842. But why link such disparate creatures—a carnivore with two herbivores, one now viewed as a tall, upright, bipedal duckbill, the other as a low, squat, four-footed, armored ankylosaur? In part, Owen didn't appreciate the extent of the differences (though we continue to regard dinosaurs as a discrete evolutionary group, thus confirming Owen's basic conclusion). For example, he didn't recognize the bipedality of some dinosaurs, and therefore reconstructed all three genera as four-footed creatures.

Owen presented three basic reasons for proposing his new group. First, the three genera share the most obvious feature that has always set our primal fascination with dinosaurs: gigantic size. But Owen knew perfectly well that similarity in bulk denotes little or nothing about taxonomic affinity. Several marine reptiles of the same age were just as big, or even bigger, but Owen did not include them among dinosaurs (and neither do we today).

Moreover, Owen's anatomical analysis had greatly reduced the size estimates for dinosaurs (though these creatures remained impressively large). Mantell had estimated up to one hundred feet in length for *Iguanodon,* a figure reduced to twenty-eight feet by Owen. In the 1844 edition of his *Medals of Creation* (only two years after Owen's shortening, thus illustrating the intensity of public interest in the subject), Mantell capitulated and excused himself for his former overestimate in the maximally exculpatory passive voice:

> In my earliest notices of the *Iguanodon* . . . an attempt was made to estimate the probable magnitude of the original by instituting a comparison between the fossil bones and those of the Iguana.

But the modern iguana grows short legs, splayed out to the side, and a long tail. The unrelated dinosaur *Iguanodon* had very long legs by comparison (for we now recognize the creature as bipedal), and a relatively shorter tail. Thus, when Mantell originally estimated the length of *Iguanodon* from very incomplete material consisting mostly of leg bones and teeth, he erred by assuming the same proportions of legs to body as in modern iguanas, and by then appending an unknown tail of greatly extended length.

Second, and most importantly, Owen recognized that all three genera share a set of distinct characters found in no other fossil reptiles. He cited many technical details, but focused on the fusion of several sacral vertebrae to form an unusually strong pelvis—an excellent adaptation for terrestrial life, and a feature long known in *Megalosaurus,* but then only recently affirmed for *Iguanodon,* thus suggesting affinity. Owen's first defining sentence in his section on dinosaurs (pages 102–3 of his 1842 report) emphasizes this shared feature: "This group, which includes at least three well-established genera of Saurians, is characterized by a large sacrum composed of five anchylosed [fused] vertebrae of unusual construction."

Third, Owen noted, based on admittedly limited evidence, that dinosaurs might constitute a complete terrestrial community in themselves, not just a few oddball creatures living in ecological corners or backwaters. The three known genera included a fierce carnivore, an agile herbivore, and a stocky armored herbivore—surely a maximal spread of diversity and ecological range for so small a sample. Perhaps the Mesozoic world had been an Age of Dinosaurs (or rather, a more inclusive Age of Reptiles, with dinosaurs on land, pterodactyls in the air, and ichthyosaurs, plesiosaurs, and mosasaurs in the sea). In this view, dinosaurs became the terrestrial component of a coherent former world dominated by reptiles.

Owen then united all his arguments to characterize dinosaurs in a particularly flattering way that justified his etymological choice of "fearfully great." In short, Owen depicted dinosaurs not as primitive and anatomically incompetent denizens of an antediluvian world, but rather as uniquely sleek, powerful, and well-designed creatures—mean and lean fighting and eating machines for a distinctive and distinguished former world. Owen emphasized this central point with a striking rhetorical device, guaranteed to attract notice and controversy: he compared the design and efficiency of dinosaurs with modern (read superior) mammals, not with slithery and inferior reptiles of either past or present worlds.

Owen first mentioned this argument right up front, at the end of his opening paragraph (quoted earlier in this essay) on the definition of dinosaurs:

> The bones of the extremities are of large proportional size, for Saurians. . . . [They] more or less resemble those of the heavy pachydermal Mammals, and attest . . . the terrestrial habits of the species.

Owen then pursues this theme of structural and functional (not genealogical) similarity with advanced mammals throughout his report—as, for example, when he reduces Mantell's estimate of dinosaurian body size by comparing their leg bones with the strong limbs of mammals, attached under the body for maximal efficiency in locomotion, and not with the weaker limbs of reptiles, splayed out to the side and imposing a more awkward and waddling gait. Owen writes:

> The same observations on the general form and proportions of the animal [*Iguanodon*] and its approximation in this respect to the Mammalia, especially to the great extinct Megatherioid [giant ground sloth] or Pachydermal [elephant] species, apply as well to the *Iguanodon* as to the *Megalosaurus.*

Owen stresses this comparison again in the concluding paragraphs of his report (and for a definite theoretical purpose embodying the theme of this essay). Here Owen speaks of "the Dinosaurian order, where we know that the Reptilian type of structure made the nearest approach to Mammals." In a final footnote (and the very last words of his publication), Owen even speculates—thus anticipating an unresolved modern debate of great intensity—that the efficient physiology of dinosaurs invites closer comparison with warm-blooded mammals than with conventional cold-blooded modern reptiles:

> The Dinosaurs, having the same thoracic structure as the Crocodiles, may be concluded to have possessed a four-chambered heart; and, from their superior adaptation to terrestrial life, to have enjoyed the function of such a highly-organized center of circulation in a degree more nearly approaching that which now characterizes the warm-blooded Vertebrata.

When we contrast artistic reconstructions made before and after Owen's report, we immediately recognize the dramatic promotion of dinosaurs from ungainly, torpid, primeval reptilian beasts to efficient and well-adapted creatures more comparable with modern mammals. We may gauge this change by comparing reconstructions of *Iguanodon* and *Megalosaurus* before and after Owen's report. George Richardson's *Iguanodon* of 1838 depicts a squat, elongated creature, presumably relegated to a dragging, waddling gait while moving on short legs splayed out to the side (thus recalling God's curse upon the serpent after a portentous encounter with Eve: "upon thy belly shalt thou go, and dust shalt thou eat all the days of thy life"). But Owen, while wrongly interpreting all dinosaurs as four-footed, reconstructed them as competent and efficient runners, with legs held under the body in mammalian fashion. Just compare

Richardson's torpid dinosaur of 1838 with an 1867 scene of *Megalosaurus* fighting with *Iguanodon,* as published in the decade's greatest work in popular science, Louis Figuier's *The World Before the Deluge.*

Owen also enjoyed a prime opportunity for embodying his ideas in (literally) concrete form, for he supervised Waterhouse Hawkins's construction of the first full-sized, three-dimensional dinosaur models—built to adorn the reopening of the Crystal Palace at Sydenham in the early 1850s. (The great exhibition hall burned long ago, but Hawkins's dinosaurs, recently repainted, may still be seen in all their glory—less than an hour's train ride from central London.) In a famous incident in the history of paleontology, Owen hosted a New Year's Eve dinner on December 31, 1853, *within* the partially completed

An 1838 (top) compared with an 1867 (bottom) reconstruction of dinosaurs, to show the spread of Owen's view of dinosaurs as agile and active.

model of *Iguanodon*. Owen sat at the head of the table, located within the head of the beast; eleven colleagues won coveted places with Owen inside the model, while another ten guests (the Victorian version of a B-list, I suppose) occupied an adjoining side table placed outside the charmed interior.

I do not doubt that Owen believed his favorable interpretation of dinosaurs with all his heart, and that he regarded his conclusions as the best reading of available evidence. (Indeed, modern understanding places his arguments for dinosaurian complexity and competence in a quite favorable light.) But scientific conclusions—particularly when they involve complex inferences about general worldviews, rather than simple records of overtly visible facts—always rest upon motivations far more intricate and tangled than the dictates of rigorous logic and accurate observation. We must also pose social and political questions if we wish to understand why Owen chose to name dinosaurs as fearfully great: who were his enemies, and what views did he regard as harmful, or even dangerous, in a larger than purely scientific sense?

When we expand our inquiry in these directions, the ironic answer that motivated this essay rises to prominence as the organizing theme of Owen's deeper motivations: he delighted in the efficiency and complexity of dinosaurs—and chose to embody these conclusions in a majestic name—because dinosaurian competence provided Owen with a crucial argument against the major evolutionary theory of his day, a doctrine that he then opposed with all the zeal of his scientific principles, his conservative political beliefs, and his excellent nose for practical pathways of professional advance.

I am not speaking here of the evolutionary account that would later bear the name of Darwinism (and would prove quite compatible with Owen's observations about dinosaurs)—but rather of a distinctively different and earlier version of "transmutation" (the term then generally used to denote theories of genealogical descent), best described as the doctrine of "progressionism." The evolutionary progressionists of the 1840s, rooting their beliefs in a pseudo-Lamarckian notion of inherent organic striving for perfection, looked upon the fossil record as a tale of uninterrupted progress within each continuous lineage of organisms. Owen, on the contrary, viewed the complexity of dinosaurs as a smoking gun for annihilating such a sinister and simplistic view.

Owen opposed progressionistic evolution for a complex set of reasons. First of all, this opinion endeared him to his patrons, and gave him leverage against his enemies. William Buckland, Owen's chief supporter, had used his Bridgewater Treatise of 1836 to argue against evolutionary progressionism by citing the excellent design of ancient beasts that should have been crude and primitive by virtue of their primeval age. Buckland invoked both *Megalosaurus* and *Iguanodon* to advance this argument (though these genera had not yet been designated as dinosaurs).

In his preface, Buckland announced an intention to show that "the phenomena of Geology are decidedly opposed" to "the derivation of existing systems of organic life . . . by gradual transmutation of one species into another." He then argued that the superb design of ancient organisms proved the constant superintendence of a loving deity, rather than a natural process of increasing excellence from initial crudity to current complexity. According to Buckland, the superb design of giant Mesozoic reptiles "shows that even in those distant eras, the same care of the common Creator, which we witness in the mechanism of our own bodies . . . was extended to the structure of creatures, that at first sight seem made up only of monstrosities." He then inferred God's direct benevolence from the excellent adaptation of *Iguanodon* teeth to a herbivorous lifestyle: we cannot "view such examples of mechanical contrivance, united with so much economy of expenditure . . . without feeling a profound conviction that all this adjustment has resulted from design and high intelligence." In his 1842 report, Owen dutifully quoted all these key statements about dinosaurian excellence as proof of God's commanding love and wisdom, while copiously citing and praising Buckland as his source of insight.

This defense of natural theology and attack upon evolutionary progressionism also positioned Owen well against his enemies. Owen denigrated the amateur and bucolic Mantell as much as he revered his urbane professional patron Buckland, but this irrelevant spat centered on social rather than ideological issues. In London, Owen faced one principal enemy at this early stage of his career, when advance to domination seemed most precarious—the newly appointed professor of zoology at University College London: Robert E. Grant.

Grant (1793–1874) ended up in disgrace and poverty for reasons that remain unresolved and more than a little mysterious.* But, in the late 1830s, Grant (who had just moved south from Edinburgh) enjoyed prominence as a newly leading light in London's zoological circles. He also became Owen's obvious and only rival for primacy. Grant had published an excellent and highly respected series of papers on the biology and classification of "lower" invertebrates, and he held advantages of age and experience over Owen.

But Grant was also a political radical, a man of few social graces, and—more relevantly—the most prominent public supporter of evolutionary progressionism in Great Britain. (In a wonderful tale for another time—and a primary illustration

*See the excellent intellectual and "forensic" work of historian Adrian Desmond on this vexatious question. Incidentally, my own inspiration for this essay began with an invitation to speak at

for the sociological principle of six degrees of separation, and the general doctrine of "what goes 'round comes 'round"—Charles Darwin had spent an unhappy student year at Edinburgh before matriculating at Cambridge. As the only light in this dark time, Darwin became very close to Grant, who must be regarded as his first important academic mentor. Of course, Darwin knew about evolution from general readings (including the works of his own grandfather Erasmus, whom Grant also much admired), and Grant's Lamarckism stood in virtual antithesis to the principle of natural selection that Darwin would later develop. But the fact remains that Darwin first learned about evolution in a formal academic setting from Grant. The mystery of Grant only deepens when we learn that the impeccably generous and genial Darwin later gave Grant such short shrift. He apparently never visited his old mentor when the two men lived in London, literally at a stone's throw of separation, after Darwin returned from the *Beagle* voyage. Moreover, Darwin's autobiography, written late in his life, contains only one short and begrudging paragraph about Grant, culminating in a single statement about Grant's evolutionism: "He one day, when we were walking together, burst forth in high admiration of Lamarck and his views on evolution. I listened in silent astonishment, and as far as I can judge without any effect on my mind."

Grant represented a threat and a power in 1842, and Owen used his antievolutionary argument about dinosaurs as an explicit weapon against this archrival. The nastiest statement in Owen's report on fossil reptiles records his unsubtle skewering of Grant's evolutionary views:

> Does the hypothesis of the transmutation of species, by a march of progressive development occasioning a progressive ascent in the organic scale, afford any explanation of these surprising phenomena? . . . A slight survey of organic remains may, indeed, appear to support such views of the origin of animated species; but of no stream of science is it more necessary, than of Paleontology, to "drink deep or taste not."

To illustrate the supposed superficiality and ignorance behind such false arguments for evolution, Owen then cites only an 1835 paper by Grant—thus clearly identifying the target of his jibes. Could anyone, moreover, make a more dismissive and scurrilous statement about a colleague than Owen's rejection of Grant by Alexander Pope's famous criterion (the subject of chapter 11):

University College in a celebration to honor the reopening of Grant's zoological museum. In reading about Grant, and developing considerable sympathy for his plight, I naturally extended my research to his enemy Owen, to dinosaurs, and ultimately to this essay.

A little learning is a dangerous thing;
Drink deep, or taste not the Pierian spring.

Owen's concluding section cites several arguments to buttress his antitransmutationist message—but dinosaurs take the stand as his star and culminating witnesses. Owen, following a standard formulation of his generation, not a unique insight or an idiosyncratic ordering, makes a primary distinction between an evolutionary version of progressionism—where each lineage moves gradually, inexorably, and unidirectionally toward greater complexity and increasing excellence of design—and the creationist style of progressionism espoused by Buckland and other natural theologians—where God creates more complex organisms for each new geological age, but the highest forms of one period do not evolve into dominant creatures of the next age.

(To rebut the obvious objection that God could then be viewed as a bumbler who couldn't get things right at the outset, and then had to use all of geological time for practice runs, Buckland and company—including Owen in the final paragraphs of his 1842 report—argued that God always creates organisms with optimal adaptations for the environments of each geological period. But these environments change in a directional manner, requiring a progressive advance of organic architecture to maintain a level of proper adaptation. Specifically, Buckland argued that climates had worsened through time as the earth cooled from an initially molten state. Cold-blooded torpor worked best on a hot and primitive earth, but a colder and tougher world required the creation of warm-blooded successors.)

Buckland and Owen held that the fossil record could act as an arbiter for this vital zoological debate; indeed, they both believed that paleontology might win primary importance as a science for this capacity to decide between evolutionary and creationist versions of progressionism. The two theories differed starkly in their predictions on a crucial matter: for transmutationists, each separate lineage should progress gradually and continuously through time, while the newly dominant form of each geological age should descend directly from the rulers of the last period. But for progressive creationists, an opposite pattern should prevail in the fossil record: individual lineages should show no definite pattern through time, and might even retrogress; while the dominant form of each age should arise by special creation, without ancestors and with no ties to the rulers of past ages.

With this background, we can finally grasp the central significance of Owen's decision to reconstruct dinosaurs as uniquely complicated beasts, more comparable in excellence of design with later (and advanced) mammals, than with lowly reptiles of their own lineage. Such dinosaurian excellence refuted transmutation and supported progressive creationism on both crucial points. First, the high status of dinosaurs proved that reptiles had degenerated through time, as the old and best—the grand and glorious *Megalosaurus* and *Iguanodon*—gave way to the later and lowlier snakes, turtles, and lizards. Second, these highest reptiles did not evolve into the next dominant group of mammals, for small and primitive mammals had already been discovered in Mesozoic rocks that housed the remains of dinosaurs as well.

Owen clearly rejoiced (and gave thanks for the support thereby rendered to his patron Buckland) in the fearful greatness of his newly christened dinosaurs as a primary argument against the demonizing doctrine of progressive transmutation. He wrote in the concluding passages of his 1842 report:

> If the present species of animals had resulted from progressive development and transmutation of former species, each class ought now to present its typical characters under their highest recognized conditions of organization: but the review of the characters of fossil Reptiles, taken in the present Report, proves that this is not the case. No reptile now exists which combines a complicated . . . dentition with limbs so proportionally large and strong, having . . . so long and complicated a sacrum as in the order *Dinosauria*. The Megalosaurs and Iguanodons, rejoicing in these undeniably most perfect modifications of the Reptilian type, attained the greatest bulk, and must have played the most conspicuous parts, in their respective characters as devourers of animals and feeders upon vegetables, that this earth has ever witnessed in oviparious [egg-laying] and cold-blooded creatures.

Owen then closed his argument, and this entire section of his report, by using dinosaurs to support the second antitransmutationist principle as well: not only do dinosaurs illustrate a lack of progress within reptilian lineages, but they also demonstrate that higher mammals could not have evolved from dominant reptiles:

> Thus, though a general progression may be discerned, the interruptions and faults, to use a geological phrase, negative [*sic*] the notion that the progression had been the result of self-developing energies adequate to a transmutation of specific characters; but, on the con-

> trary, support the conclusion that the modifications of osteological structure which characterize the extinct Reptiles, were originally impressed upon them at their creation, and have been neither derived from improvement of a lower, nor lost by progressive development into a higher type.

As a closing fillip and small (but pretty) footnote to this argument from the public record, I can also add a previously unknown affirmation for the centrality of antievolutionism as a primary motivation in Owen's designation of dinosaurs as "fearfully great." Owen certainly stressed an antitransmutationist message in discoursing on the significance of dinosaurs. But how do we know that transmutation represented a live and general debate in the zoology of Owen's day—and not just a funny little side issue acting as a bee in Owen's own idiosyncratic bonnet? The public record does provide support for the generality. For example, the first full-scale defense of evolution written in English, the anonymously printed *Vestiges of the Natural History of Creation* (by the Scottish publisher Robert Chambers), became the literary sensation and hottest press item of 1844.

But I can add a small testimony from a personal source. Several years ago, I had the great good fortune to purchase, at a modest price before the hype of *Jurassic Park* sent dinosaur memorabilia through the financial roof, Buckland's personal copy of Owen's 1842 report. (This copy, inscribed to Buckland by Owen, and signed by Buckland in two places, bears the incorrect date of 1841, and must therefore belong to the original lot of twenty-five, printed for Owen's private distribution.) Buckland obviously read the document with some care, for he underlined many passages and wrote several marginal annotations. But the annotations follow a clear pattern: Buckland only highlights factual claims in his marginal notes; he never comments on theoretical or controverted points. Mostly, he just lists the taxonomic names of species under discussion, or the anatomical terms associated with Owen's immediate descriptions. For example, in Owen's section on dinosaurs, Buckland writes "sacrum" next to Owen's identification of fused sacral vertebrae as a defining character of dinosaurs. And he writes "28 feet" next to Owen's defense of this smaller length for *Iguanodon*.

Buckland breaks this pattern only once—to mark and emphasize Owen's discussion of dinosaurs as an argument against transmutation of species. Here, on page 196, Buckland writes the word "transmutation" in the margin—his only annotation for a theoretical point in the entire publication. Moreover, and in a manner that I can only call charming in our modern age of the 3M

Post-it, Buckland cut out a square of white paper and fastened it to page 197 by a single glob of marginal glue, thus marking Owen's section on evolution by slight projection of the square above the printed page when the book lies closed. Finally, Buckland again wrote the single word "Transmutation" on a loose, rectangular slip of paper that he must have inserted as a bookmark into Owen's report. Buckland, one of the leading English geologists of his time, evidently regarded Owen's discussion of evolution as the most important theoretical issue addressed by the giant reptiles that Buckland had first recognized and that Owen had just named as dinosaurs.

Evolution must be a genuinely awful and absolutely terrible truth if Owen felt compelled to employ the most fearfully great of all fossil creatures in an ultimately vain attempt to refute this central principle of life's history, the source of all organic diversity from *Megalosaurus* to Moses, from *Iguanodon* to the "lowly" infusorians residing inside those zoological oddballs who learn to name dinosaurs and strive to contemplate the great I Am.

10

Second-Guessing the Future

FROM ANONYMOUS VICE-PRESIDENTS TO NAMELESS PALOOKAS, a special kind of opprobrium seems to haunt those who finish second—close but no cigar, in an old cliché. I once met "Two Ton Tony" Galento in a bar in upstate New York, a pitiful figure as an old man, still cadging drinks in exchange for the true story of his moment of glory: when he knocked Joe Louis down before losing their fight for the heavyweight championship. And just consider the stereotype of the sidekick—old, fat, foolish, and in servitude—from Gabby Hayes and Andy Devine in the quintessential epic of our pop culture, to Leporello and Sancho Panza in the literary world. (Strong and noble sidekicks like Tonto get cast as "ethnics" to advertise their secondary rank by another route, now happily—or at least hopefully—fading from the collective consciousness of white America.)

Second in time fares no better than second in status. I was, at first, surprised by a statement that made perfect sense once I punctured the apparent paradox. A composer friend told me that he could easily obtain funding for a premiere performance of any new work—as special grants and scholarships abound for such a noble purpose. A philanthropist who truly loved music, he told me, would endow the most unprofitable and unfashionable of all genres: *second* performances of new works.

I recently had the privilege of speaking with Larry Doby, one of the toughest, most courageous, and most admirable men I have ever met. But how many readers recognize his name? We all know Jackie Robinson, who came first; Larry Doby was the second black player in Major League Baseball (and first in the American League). We all recognize the tune when Rodolfo grasps Mimi's cold little hand in Puccini's *La Bohème,* first performed in 1896. But how many people know that Leoncavallo (who had scored the hit of 1892 with *I Pagliacci*) also wrote an opera with the same title (and tale) in 1897?

I can think of only one second finisher who became more famous (at least among Anglophones) than the victor—but only for special circumstances of unusual heroism in death, mingled with a dose of overextended British patriotism: Robert Scott, who reached the South Pole on January 18, 1912, only to find that Roald Amundsen had beat him to the bottom by a full month. Confined to a tent by a blizzard, and just eleven miles from his depot, Scott froze to death, leaving a last journal entry that has never been matched in all the annals of British understatement, and that, I confess, still brings tears to my eyes: "It seems a pity, but I do not think I can write more."

In my parish, the dubious (and admittedly somewhat contradictory) status of most famous second-place finisher goes without contest to Alfred Russel Wallace, who, in 1858 during a malarial fit on the Indonesian island of Ternate, devised virtually the same theory of natural selection that Darwin had developed (but never published) in 1838. In a familiar story, Wallace sent his short paper to Darwin, a naturalist he greatly admired and who, as Wallace knew, maintained a strong interest in "the species question" (though Wallace had no inkling of Darwin's particular and nearly identical theory, and probably didn't even realize that Darwin had a theory at all). Darwin, in understandable panic, turned to his best friends, Charles Lyell and Joseph Hooker, for advice. In a resolution known to later history as the "delicate arrangement," Darwin's friends made a joint presentation to the Linnaean Society of London in July 1858. At this meeting, they read both Wallace's paper and some unpublished letters and manuscripts by Darwin, illustrating his earlier authorship of the same idea.

Conspiracy theorists always stand at the ready, and several salvos have been launched for this particular episode, but to no avail or validity, in my judgment. Yes, Wallace was never asked (being quite incommunicado, half a world away, while time did press). Yes, Darwin was wealthy and well established; Wallace, poor, younger, and struggling for livelihood and reputation. (But why, then, grant him equal billing with Darwin for a joint presentation of unpublished results?) No, I think that, as usual (and unfortunately for the cause of a good tale), the more boring resolution of ordinary decency applies.

"Delicate arrangement" describes the result quite accurately: a fair solution to a tough problem. Darwin held legitimate priority, and he had not been shilly-shallying, or resting on old claims and laurels. He had been diligently working on his evolutionary views and had already, when he received Wallace's paper, finished nearly half of a much longer book on natural selection that he then abandoned (spurred no doubt by fears of further anticipations) to write the shorter "abstract" known to the world as *The Origin of Species* (a pretty hefty book of 490 pages), published in 1859.

Wallace, at least, never complained, and seemed to feel honored that his exercise of an evening had been thus linked with Darwin's long effort. (I do not, of course, base this claim on Wallace's public pronouncements, where his secondary status to Darwin would have precluded any overt expression of bitterness. Rather, in his truly voluminous private jottings, letters, and conversations, Wallace never expressed anything but pleasure at Darwin's willingness to share at least partial credit.)

I do not, however, deny the usual assessment of Wallace as a man trammeled by meager circumstances and dogged by hard luck. He spent several youthful years of difficult and dangerous fieldwork in the Amazon, only to lose all his specimens in a shipwreck that nearly ended his own life as well. Wallace did not despair, but quickly set sail in the other direction, and spent several years engaged in similar work around the Malay Archipelago, where he took second place in the greatest biological discovery in history. He grew up in poverty (in a family of middle-class social status but much lower means), and while comfortable enough during most of his adult life, he never accumulated adequate resources to reach his true goal: doing science without impediment, and without needing to live by his own wits as a writer and lecturer. (A government pension, secured for Wallace by Darwin and his friends—perhaps partly to assuage a tinge of guilt—didn't hurt, but didn't guarantee solvency either.)

Because Wallace lived a long time (1823–1913), wrote copiously both for his bread and from his convictions, and held a variety of passionate and quirky views, he left us a vast legacy of varied content and quality. He campaigned ardently for the right and the just according to his idiosyncratic standards, and he fought valiantly for a set of causes usually deemed "cranky," both in his own

The illustrations for this chapter were published in France in the late nineteenth century and represent predictions for twentieth-century achievements.

time and today—including phrenology and spiritualism (where he nearly came to blows with skeptics like Darwin and Huxley)—and against vaccination, which he called "one of the foulest blots on the civilization of the 19th century." His politics defy simple characterization, but generally fall into a camp that might be labeled as democratic socialism of a Fabian bent, but spiced by utter devotion to a few favored causes that did not rank high on most people's list of indispensable reforms.

I have often called upon Wallace's large body of work for essays in these books, both for his wisdom (in debunking Percival Lowell's ideas on Martian canal builders), and for his crankiness (in claiming virtual proof for the proposition that, throughout the entire universe, no planet but the earth could house intelligent life). But now, for the first time, I invoke Wallace proactively, and after considerable patience in waiting for the appropriate moment.

An impassioned author, approaching a public turning point at the height of his own supposed wisdom and maturity, could scarcely resist such a temptation for proclamation. The turnings of our centuries may bear no relationship to any natural cycle in the cosmos. (I label such passages as "precisely arbitrary" in the subtitle to my previous book, *Questioning the Millennium*.) But we construe such artificial transitions as occasions for taking stock, especially at the centurial boundaries that have even generated their own eponymous concept of cyclical *Angst*—the *fin de siècle* (end of century) phenomenon. (The forthcoming millennium might provoke an even greater burst, but we have too little experience for any prediction. I am at least amused by a diminution in the quality of anxiety for the two documented transitions: last time around, Europe feared all the gory prophecies of Armageddon, as recorded in Revelation 20. For this second experience in Western history, we focus our worries on what might happen if computers misread the great turning as a recursion to the year 1900.)

Thus, Alfred Russel Wallace could not let the nineteenth century expire without presenting his summation, his evaluation, and his own predictions to the world. He published *The Wonderful Century: Its Successes and Failures* in 1898, and I have been waiting for several years to celebrate the hundredth anniversary of this book near the dawn of our own new millennium.* I saved my remarks for this forum of evolutionary essays, both because Wallace plays a major role on this particular stage, and because the genre of fin-de-siècle summations includes two linked and distinctive themes that have served as linchpins for these essays: the relationship between science and society (an unavoidable centerpiece in assessing the nineteenth century, with its technologically inspired industrial and colonial expansions), and the unpredictability of evolutionary and social futures (ironically, the theme that ultimately undermines this entire genre of summing up the past in hopes of securing a better future).

Wallace presents a simple thesis as the foundation for his epitome of the nineteenth century—a standard view about the relation of science to society, stated in the context of a particular time. Science, Wallace argues, has made unprecedented gains, largely expressed as technological advance (at least in terms of impacts upon everyday life), but this progress has been blunted, if not perverted, by our failure to make any moral improvements, especially as expressed in the alleviation of social inequities. Thus, and ironically, the progress of science, however bursting with potential for social improvement as well, has actually operated to increase the sum total of human misery.

Wallace opens with a statement of his thesis:

> The present work is not in any sense a history, even on the most limited scale. It may perhaps be termed an appreciation of the century—of what it has done, and what it has left undone. . . . A comparative estimate of the number and importance of these [material and intellectual] achievements leads to the conclusion that not only is our century superior to any that have gone before it, but that it

*I wrote and first published this essay in 1998.

> may be best compared with the whole preceding historical period. It must therefore be held to constitute the beginning of a new era in human progress. But this is only one side of the shield. Along with these marvelous Successes—perhaps in consequence of them—there have been equally striking Failures, some intellectual, but for the most part moral and social. No impartial appreciation of the century can omit a reference to them; and it is not improbable that, to the historian of the future, they will be considered to be its most striking characteristic.

In his first, and shorter, section on scientific and technological progress, Wallace even tries to quantify the relative value of nineteenth-century achievements, reaching the conclusion that this single century had surpassed the summation of all previous human history in weight of accumulated progress:

> In order to estimate its [the nineteenth century's] full importance and grandeur—more especially as regards man's increased power over nature, and the application of that power to the needs of his life today, with unlimited possibilities in the future—we must compare it, not with any preceding century, or even with the last millennium, but with the whole historical period—perhaps even with the whole period that has elapsed since the stone age.

The chapters of this first part then detail the major inventions, spurred by advancing science, that brought such great potential improvement to nineteenth-century life: control of fire (with wide-ranging implications from steam engines to generating plants), labor-saving machinery, transportation, communication, and lighting (culminating in the incandescent bulb). Wallace's examples often combine charm with insight (as we recall, from yet a century further on, the different lives of not-so-distant forebears). For example, Wallace writes of his own childhood:

> The younger generation, which has grown up in the era of railways and of ocean-going steamships, hardly realize the vast change which we elders have seen. . . . Even in my own boyhood the wagon for the poor, the stage coach for the middle class, and the post-chaise for the wealthy, were the universal means of communication, there being only two short railways then in existence. . . . Hundreds of

> four-horse mail and stage coaches, the guards carrying horns or bugles which were played while passing through every town or village, gave a stir and liveliness and picturesqueness to rural life which is now almost forgotten.

I confess to a personal reason for intrigue with Wallace's best example for regarding the nineteenth century as exceeding all previous history in magnitude of technological improvement: the trip from London to York, he states, took less time during the Roman occupation than in 1800, just before the advent of railroads—for the Romans built and maintained better roads, and horses moved no faster in 1800 than in A.D. 300. (I am amused by the analogous observation that rail travel on my frequent route between New York and Boston has slowed during the last century. A nineteenth-century steam engine could make the journey faster than Amtrak's quickest train, which now runs by electricity from New York to New Haven, but must then lose substantial time in switching engines for the diesel run on a nonelectrified route from New Haven to Boston. Yes, they tell us, vast improvement and full electrification lie just around the temporal corner. But how long, oh Lord, how long!)

In reading Wallace's examples, I also appreciated the numerous reminders of the central principle that all truly creative invention must be tentative and flexible, for many workable and elegant ideas will be quickly superseded—as in this temporary triumph for news over the newly invented telephone:

> Few persons are aware that a somewhat similar use of the telephone is actually in operation at Buda Pesth [*sic* for Budapest, a city then recently amalgamated from two adjoining towns with Wallace's separate names] in the form of a telephonic newspaper. At certain fixed hours throughout the day a good reader is employed to send definite classes of news along the wires which are laid to subscribers' houses and offices, so that each person is able to hear the particular items he desires, without the delay of its being printed and circulated in successive editions of a newspaper. It is stated that the news is supplied to subscribers in this way at little more than the cost of a daily newspaper, and that it is a complete success.

But Wallace's second and longer section then details the failures of the nineteenth century, all based on the premise that moral stagnation has perverted the application of unprecedented scientific progress:

> We of the 19th century were morally and socially unfit to possess and use the enormous powers for good or evil which the rapid advance of scientific discovery had given us. Our boasted civilization was in many respects a mere surface veneer; and our methods of government were not in accordance with either Christianity or civilization. This view is enforced by the consideration that all the European wars of the century have been due to dynastic squabbles or to obtain national aggrandizement, and were never waged in order to free the slave or protect the oppressed without any ulterior selfish ends.

Wallace then turns to domestic affairs, with the damning charge that our capitalist system has taken the wealth accrued from technological progress, and distributed the bounty to a few owners of the means of production, while actually increasing both the absolute and relative poverty of ordinary working people. In short, the rich get richer and the poor get poorer:

> One of the most prominent features of our century has been the enormous and continuous growth of wealth, without any corresponding increase in the well-being of the whole people; while there is ample evidence to show that the number of the very poor—of those existing with a minimum of the bare necessities of life—has enormously increased, and many indications that they constitute

> a larger proportion of the whole population than in the first half of the century, or in any earlier period of our history.

At his best, Wallace writes with passion and indignation, as in this passage on preventable industrial poisoning of workers:

> Let every death that is clearly traceable to a dangerous trade be made manslaughter, for which the owners . . . are to be punished by imprisonment. . . . and ways will soon be found to carry away or utilize the noxious gases, and provide the automatic machinery to carry and pack the deadly white lead and bleaching powder; as would certainly be done if the owners' families, or persons of their own rank of life, were the only available workers. Even more horrible than the white-lead poisoning is that by phosphorus, in the match-factories. Phosphorus is not necessary to make matches, but it is a trifle cheaper and a little easier to light (and so more dangerous), and is therefore still largely used; and its effect on the workers is terrible, rotting away the jaws with the agonizing pain of cancer followed by death. Will it be believed in future ages that this horrible and unnecessary manufacture, the evils of which were thoroughly known, was yet allowed to be carried on to the very end of this century, which claims so many great and beneficent discoveries, and prides itself on the height of civilization it has attained?

Wallace offers few specific suggestions for a new social order, but he does state a general principle:

> The capitalists as a class have become enormously richer. . . . And so it must remain till the workers learn what alone will save them, and take the matter into their own hands. The capitalists will consent to nothing but a few small ameliorations, which may improve the condition of select classes of workers, but will leave the great mass just where they are.

I doubt that Wallace harbored any muscular or martial fantasies about armed revolt sweeping through the streets of London, with the apostles of a new and better world, himself included, leading a vanguard, rifles held high. Wallace was far too gentle a man even to contemplate such a style of renewal. At most, he looked to electoral reform and unionization as means for workers to take "the matter into their own hands." His final chapter, entitled "The Remedy for Want," goes little beyond a naive proposal for free bread on demand, financed by a voluntary (albeit strongly suggested) governmental tax upon people with the highest incomes.

Wallace's summary of the nineteenth century—a steady inexorability of technological progress derailed by failure of our moral and social sensibilities to keep pace—underscores the second evolutionary theme of this essay, while undermining the entire genre of fin-de-siècle (or millennium) summations: the unpredictability of human futures, and the futility of thinking that past trends will forecast coming patterns. The trajectory of technology might offer some opportunity for prediction—as science moves through networks of implication, and each discovery suggests a suite of following steps. (But even the "pure" history of science features unanticipated findings, and must also contend with nature's stubborn tendency to frustrate our expectations—factors that will cloud anyone's crystal ball.) Moreover, any forecast about the future must also weigh the incendiary instability generated by interaction between technological change and the weird ways of human conduct, both individual and social. How, then, can the accidents that shaped our past give any meaningful insight into the next millennium?

I think that the past provides even dimmer prospects for prediction than Wallace's model of history implies—for another destabilizing factor must be added to Wallace's claim for discordance between technological and moral change. Wallace missed the generality of an important pattern in nature because he remained so committed to Lyellian (and Darwinian) gradualism as the designated way of life on earth. His book devotes an entire chapter (in the first section on scientific progress) to arguing that the replacement of catastrophism by uniformitarian geology—the notion that major features of the earth's history and topography "are found to be almost wholly due to the slow action of the most familiar everyday causes" and should not be "almost always explained as being due to convulsions of nature"—"constitutes one of the great philosophical landmarks of the 19th century."

Wallace knew that the discordance of technological and moral change could produce catastrophic disruption in human history, but he viewed such a result as exceptional among the ways of nature, and not subject to generalization. Now that our modern sensibilities have restored catastrophism as an important option (though not an exclusive pattern) for nature as well, this theme gains ground as a powerful argument against predictability. Not only as an anomaly of human history, but also as a signature of nature, pasts can't imply futures because a pattern inherent in the structure of nature's materials and laws—"the great asymmetry" in my terminology—too often disrupts an otherwise predictable unfolding of historical sequences.

Any complex system must be constructed slowly and sequentially, adding steps one (or a few) at a time, and constantly coordinating along the way. But the same complex systems, once established, can be destroyed in a tiny fraction of the necessary building time—often in truly catastrophic moments—thus engendering the great asymmetry between building up and tearing down. A day of fire destroyed a millennium of knowledge in the library of Alexandria, and centuries of building in the city of London. The last blaauwbock of southern Africa, the last moa of New Zealand, perished in a momentary blow or shot from human hands, but took millions of years to evolve.

The discordance between technological and moral advance acts as a destabilizing factor to feed the great asymmetry, and prevents us from extrapolating past trends into future predictions—for we never know when and how the ax of the great asymmetry will fall, sometimes purging the old to create a better world by revolution, but more often (I fear) simply cutting a swath of destruction, and requiring a true rebirth from the ashes of old systems (as life has frequently done—in a wondrously unpredictable way—following episodes of mass extinction).

Thus, I am even less sanguine than Wallace about possibilities for predicting the future—even though I think that he overstated his case in an important way. I don't fully agree with Wallace's major premise that technology has progressed while morality stagnated. I rather suspect that general levels of morality have improved markedly as well, at least during the last millennium of Western history—though I don't see how we could quantify such a claim. In most of the world, we no longer keep slaves, virtually imprison women, mock the insane, burn witches, or slaughter rivals with such gleeful abandon or such unquestioned feelings of righteousness. Rather, our particular modern tragedy—and our resulting inability to predict the future—resides largely in the great asymmetry, and the consequential, if unintended, power of science to enhance the effect. I suspect that twenty Hitlers ruled over small groups of Europeans a thousand years ago. But what could such petty monsters accomplish with bows and arrows, battering rams, and a small cadre of executioners? Today, one evil man can engineer the murder of millions in months.

Finally, a fascinating effect of scale defeats all remaining hope for meaningful predictability. Yes, if one stands way, way back and surveys the history of

human technology, I suppose that one might identify a broad form of sensible order offering some insight into future possibilities. The invention of agriculture does imply growth in population and construction of villages; gunpowder does move warfare away from the besieging of walled cities; and computers must exert some effect upon printed media. Unless the great asymmetry wipes the slate clean (or even frees the earth from our presence entirely), some broad patterns of technological advance should be discernible amidst all the unpredictable wriggles of any particular moment.

Yes, but almost all our agonized questions about the future focus upon the wriggles, not the broader patterns of much longer scales. We want to know if our children will be able to live in peace and prosperity, or if the Statue of Liberty will still exist to intrigue (or bore) our grandchildren on their school trips, or to greet yet another wave of immigrants. At most, we ask vague and general questions about futures not really very distant, and not truly very different, from what we already know or suspect.

Just consider the most widely discussed pattern of human history since the invention of writing: the rise, spread, and domination of the European world, thanks largely to the auxiliary technologies of gunpowder and navigation. Traditions of Western explanation, largely self-serving of course, have focused upon two successive causes—strikingly different claims to be sure, but strangely united in viewing European domination as predictable, if not foreordained.

The first, as old as our lamentable self-aggrandizement, simply trumpets the inherent superiority of European people, a claim made even uglier in the last few centuries by grafting the phony doctrine of scientific racism upon old-fashioned xenophobia. The second—arising largely from a desire to reject the falsity and moral evil of racism, while still viewing history as predictably sensible—holds that people are much of a muchness throughout the world, but that certain climates, soils, and environments must inspire technological advance, and European people just happened to live in the right place.

This second argument holds much merit, and almost has to be valid at a scale of explanation that steps way back and looks at broadest patterns. Indeed, no other explanation in the determinist mode makes any sense, once we recognize the multitude of recent genetic studies that reveal only trivial differences among human groups, based on an enormous weight of shared attributes and the great variability existing *within* each of our groups.

But again, I ask most readers of this essay (originally published in a Western land and language, and initially read mostly by people of European descent) to look into their guts and examine the basis of their question: are you really asking about an admittedly broad inevitability based on soils and latitudes, or are you wondering about a wriggle lying within the realm of unpredictability? I suspect that most of us are really asking about wriggles, but looking at the wrong scale and thinking about predictability.

Yes, complex technology probably had to emerge from mid-latitude people living in lands that could support agriculture—not from Eskimos or Laplanders in frozen terrains with limited resources, and not from the hottest tropics, with vegetation too dense to clear, and a burden of disease too great to

bear. But which mid-latitude people? Or to be more honest (and for the majority of Anglophones who read this essay in its original form), why among people of my group and not of yours?

In honest and private moments, I suspect that most readers of European descent regard the spread of European domination as a sensible and predictable event, destined to happen again if we could rewind time's tape, say to the birth of Jesus, and let human history unroll on a second and independent run. But I wouldn't bet a hoplite's shield or a Frenchman's musket on a rerun with European domination. The little wriggles of a million "might have beens" make history, not the predictabilities of a few abstract themes lying far from our concerns in a broad and nebulous background.

Can we really argue that Columbus's caravels began an inevitable expansion of one kind of people? Surely not when the great Chinese admiral Zheng He (rendered as Cheng Ho in a previously favored system of transliteration), using a mariner's compass invented by his people, led seven naval expeditions as far as the shores of eastern Africa between 1405 and 1433. Some of Zheng He's ships were five times as long as a European caravel, and one expedition may have included as many as sixty-two ships carrying nearly 28,000 men.

To be sure, Zheng He sailed for the Yung-lo emperor, the only ruler who ever favored such expansionist activities during the Ming dynasty. His successors suppressed oceanic navigation and instituted a rigidly isolationist policy. (I also understand, though I can claim no expertise in Chinese history, that Zheng He's voyages must be viewed more as tributary expeditions for glorifying the

emperor, than as harbingers of imperialistic expansion on Western models. Incidentally, as further evidence for our fascination with differences, I have never read a document about Zheng He that proceeds past the first paragraph before identifying the great admiral as both a Muslim and a eunuch. I could never quite fathom the relevance, for captains don't navigate with their balls, while we know that court eunuchs played a major role throughout Chinese imperial history.)

In any case, suppose that Chinese history had unfolded a bit differently? Suppose that the successors of the Yung-lo emperor had furthered, rather than suppressed, his expansionist policies? Suppose that subsequent admirals had joined another great Chinese invention—gunpowder as weaponry—with their unmatched naval and navigational skills to subdue and occupy foreign lands? May we not suppose that Caucasian Europe would then have become a conquered backwater?

We must also consider dramatic (and entirely believable) alternatives within Caucasian history. Has any force in human affairs ever matched the spreading power of Islam after a local origin in the sixth century A.D.? The preeminent traveler Ibn Battuta surveyed the entire Muslim world during three decades of voyaging in the mid-fourteenth century. Would any companion have bet on Christianity over Islam at that moment in history? (And how would one vote today, despite the intervening success of European doctrines?) The *Encyclopaedia Britannica* comments: "Thomas Aquinas (c. 1224–1274) might have been read from Spain to Hungary and from Sicily to Norway; but Ibn al-'Arabi (1165–1240) was read from Spain to Sumatra and from the Swahili coast to Kazan on the Volga River."

Islam came close to subduing Europe on several occasions that might easily have experienced an opposite outcome. Perhaps that Moors of Iberia never did have designs on all Europe, despite the cardboard tale we once learned in conventional Western history classes—that Islam peaked and began an inevitable decline when Charles Martel defeated the Moors at Poitiers in 732. *Britannica* remarks that "the Andalusian Muslims never had serious goals across the Pyrenees. In 732, Charles Martel encountered not a Muslim army, but a summer raiding party."

But genuine threats persisted for nearly a thousand years. If the great Timur (also known as Tamerlane), the Turkic conqueror of Samarkand, had not turned his sights toward China, and died in 1405 before his eastern move, Europe might also have fallen to his form of Islam. And the Ottoman sultans, with their trained and efficient armies, took Constantinople (now Istanbul) in 1453, and laid powerful siege to the walls of Vienna as late as 1683—a final failure that gave us the croissant as a living legacy, the breakfast roll based on the Turkish symbol of a crescent moon, and first made by Viennese bakers to celebrate their victory. (As a little footnote, remember that I have not even mentioned Attila, Genghis Khan, and several other serious threats to European domination.)

Our history could have been fashioned in a million different credible ways, and we have no adequate sense of where we are heading. But a good moral compass, combined with an intelligent use of scientific achievements, might keep us going—even prospering—for a long time by our standards (however paltry in geological perspective). We do have the resources, but can we muster both the will and judgment to hold first place in a game that can offer only possibilities, never guarantees—a game that spells oblivion for those who win the opportunity but fail to seize the moment, plunging instead into the great asymmetry of history's usual outcome?

IV

Six Little Pieces on the Meaning and Location of Excellence

Substrate and Accomplishment

11

Drink Deep, or Taste Not the Pierian Spring

MOST FAMOUS QUOTATIONS ARE FABRICATED; AFTER all, who can concoct a high witticism at a moment of maximal stress in battle or just before death? A military commander will surely mutter a mundane "Oh shit, here they come" rather than the inspirational "Don't fire until you see the whites of their eyes." Similarly, we know many great literary lines by a standard misquotation rather than an accurate citation. Bogart never said "Play it again, Sam," and Jesus did not proclaim that "he who lives by the sword shall die by the sword." Ironically, the most famous of all quotations about the cardinal subject of our mental lives—learning—bungles the line and substitutes "knowledge" for the original. So let us restore the proper word to Alexander Pope's "Essay on Criticism":

A little learning is a dangerous thing;
Drink deep, or taste not the Pierian Spring;
There shallow draughts intoxicate the brain,
And drinking largely sobers us again.

I have a theory about the persistence of the standard misquotation, "a little knowledge is a dangerous thing," a conjecture that I can support through the embarrassment of personal testimony. I think that writers resist a full and accurate citation because they do not know the meaning of the crucial second line. What the dickens is a "Pierian Spring," and how can you explain the quotation if you don't know? So you extract the first line alone from false memory, and "learning" disappears.

To begin this short essay about learning in science, I vowed to explicate the Pierian spring so I could dare to quote this couplet that I have never cited for fear that someone would ask. And the answer turned out to be joyfully accessible—a two-minute exercise involving one false lead in an encyclopedia (reading two irrelevant articles about artists named Piero), followed by a good turn to *The Oxford English Dictionary.* Pieria, this venerable source tells us, is "a district in northern Thessaly, the reputed home of the muses." Pierian therefore becomes "an epithet of the muses; hence allusively in reference to poetry and learning."

So I started musing about learning. Doesn't my little story illustrate a general case? We fear that something we want to learn will be difficult and that we will never even figure out how to find out. And then, when we actually try, the answer comes easily—with joy in discovery, for no delight can exceed the definitive solution to a little puzzle. Easy, that is, so long as we can master the tools (not everyone enjoys immediate access to *The Oxford English Dictionary;* more sadly, most people never learned how to use this great compendium or even know that it exists). Learning can be easy because the human mind works as an intellectual sponge of astonishing porosity and voracious appetite, that is, if proper education and encouragement keep the spaces open.

A commonplace of our culture, and the complaint of teachers, holds that, of all subjects, science ranks as the most difficult to learn and therefore the scariest and least accessible of all disciplines. Science may occupy the center of our practical lives, but its content remains mysterious to nearly all Americans, who must therefore take its benefits on faith (turn on your car or computer and pray that the damned thing will work) or fear its alien powers and intrusions (will my clone steal my individuality?). We suspect that public knowledge of science may be extraordinarily shallow, both because few people show any interest or familiarity for the subject (largely through fear or from assumptions of utter incompetence) and because those who profess concern have too superficial an understanding. Therefore, to invoke Pope's topsy-turvy metaphor again, Americans shun the deep drink that sobriety requires and maintain dangerously little learning about science.

But I strongly suspect that this common, almost mantralike belief among educators represents a deep and (one might almost say, given the vital importance and fragility of education) dangerous fallacy, arising as the product of a common error in the sciences of natural history, including human sociology in this case—a false taxonomy. I believe that science is wonderfully accessible, that most people show a strong interest, and that levels of general learning stand quite high (within an admittedly anti-intellectual culture overall), but that we have mistakenly failed to include the domains of maximal public learning within the scope of science. (And like Pope, I do distinguish learning, or visceral understanding by long effort and experience, from mere knowledge, which can be mechanically copied from a book.)

I do not, of course, hold that most people have developed the highly technical skills that lead to professional competence in science. But this situation prevails for any subject or craft, even in the least arcane and mathematical of the humanities. Few Americans can play the violin in a symphony orchestra, but nearly all of us can learn to appreciate the music in a seriously intellectual way. Few can read ancient Greek or medieval Italian, but all can revel in a new translation of Homer or Dante. Similarly, few can do the mathematics of particle physics, but all can understand the basic issues behind deep questions about the ultimate nature of things and even learn the difference between a charmed quark and a top quark.

For the false taxonomy, we don't restrict adequate knowledge of music to professional players; so why do we limit understanding of science to those who live in laboratories, twirl dials, and publish papers? Taxonomies are theories of knowledge, not objective pigeonholes, hatracks, or stamp albums with places preassigned. A false taxonomy based on a bogus theory of knowledge can lead us badly astray. When Guillaume Rondelet, in his classic monograph of 1555 on the taxonomy of fishes, began his list of categories with "flat and compressed fishes," "those that dwell among the rocks," "little fishes" *(pisciculi),* "genera of lizards," and "fishes that are almost round," he pretty much precluded any deep insight into the truly genealogical basis of historical order.

Millions of Americans love science and have learned the feel of true expertise in a chosen expression. But we do not honor these expressions by categorization within the realm of science, although we certainly should, for they encompass the chief criteria of detailed knowledge about nature and critical thinking based on logic and experience. Consider just a small list, spanning all ages and classes and including a substantial fraction of our population. If all these folks understood their engagement in *doing* science actively, democracy would shake hands with the academy, and we might learn to harvest a deep and widespread fascination in the service of more general education. (I thank Philip Morrison, one of America's wisest scientists and humanists, for making this argument to me many years ago, thus putting my thinking on the right track.)

1. Sophisticated knowledge about underwater ecology among tropical fish enthusiasts, mainly blue-collar males and therefore mostly invisible to professional intellectuals who tend to emerge from other social classes.

2. The horticultural experience of millions of members in thousands of garden clubs, mostly tenanted by older middle-class women.

3. The upper-class penchant for birding, safaris, and ecotourism.

4. The intimate knowledge of local natural history among millions of hunters and fishermen.

5. The astronomical learning (and experience in fields from practical lens grinding to theoretical optics) of telescope enthusiasts, with their clubs and journals.

6. The technological intuitions of amateur car mechanics, model builders, and weekend sailors.

7. Even the statistical knowledge of good poker players and racetrack touts. (The human brain works especially poorly in reasoning about probability, and I can cite no greater impediment to truly scientific thinking. But many Americans have learned to understand probability through the ultimate challenge of the pocketbook.)

8. In my favorite and clinching example, the dinosaur lore so lovingly learned by America's children. How I wish that we could quantify the mental might included in all the correct spellings of hideously complex dinosaur names among all the five-year-old children in America. Then we could truly move mountains.

Common belief could not be more ass-backward. We think that science is intrinsically hard, scary, and arcane, and that teachers can only beat the necessary knowledge, by threat and exhortation, into a small minority blessed with innate propensity. No. Most of us begin our education with an inborn love of science (which is, after all, only a method of learning the facts and principles of the natural world surrounding us, and how can anyone fail to be stirred by such an intimate subject?). This love has to be beaten *out* of us if we later fall by the wayside, perversely led to say that we hate or fear the subject. But the same love burns brightly throughout the lives of millions, who remain amateurs in the precious, literal sense of the word ("those who love") and who pursue "hobbies" in scientific fields that we falsely refuse to place within the taxonomic compass of the discipline.

And so, finally, the task of nurture and rescue falls to those people who represent what I have often called the most noble word in our language, the teachers. (*Parent* holds second place on my list; but teachers come first because parents, after an initial decision, have no further choice.) Rage (and scheme) against the dying of the light of childhood's fascination. And then emulate English literature's first instructor, the clerk of Oxenford in Chaucer's *Canterbury Tales,* the man who opened *both* ends of his mind and heart, for "gladly wolde he lerne, and gladly teche."

12

Requiem Eternal*

IN 1764, THE ENGLISH SAVANT DAINES BARRINGTON tested a visiting musical prodigy for his skills in memory, performance, composition, and improvisation. The amazed listener expressed great skepticism about his subject's stated age of eight, and wondered if father Leopold had been passing off a well-trained adult midget as his young son. Barrington therefore delayed his written account for six years until he could obtain proof, in the form of a birth certificate for Johannes Chrysostomus Wolfgangus Theophilus Mozart (later shortened to Wolfgang Amadeus by the composer himself), from an unimpeachable source: "his excellence," in Barrington's description, "Count Haslang, envoy extraordinary and minister plenipotentiary of the electors of Bavaria and Palatine."

Barrington noted that many precocious geniuses die young, and ended his article with a prayer that Mozart might live as long as

*Originally written for the pamphlet accompanying a Penguin CD of Mozart's *Requiem*.

Britain's most celebrated German import, G. F. Handel, who had died five years previously at age seventy-four. Barrington wrote: "It may be hoped that little Mozart may possibly attain to the same advanced years as Handel, contrary to the common observation that such *ingenia praecocia* are generally short lived."

Well, Mozart lived long enough to become Mozart, while failing to attain even half Handel's age. He died in 1791, at age thirty-five, with his greatest and final work, his *Requiem,* unfinished. He wrote the very last note of his life to the painfully appropriate text: *Lachrymosa dies illa*—this day full of tears. No musical composition has ever moved more people to tears, or inspired more mythological nonsense, including tales of a masked man commissioning the piece in secret, and a mysterious poisoner using the opportunity to end an opponent's life. (Peter Shaffer, of course, wove all these fictions together into a sublime play full of psychological truth—*Amadeus.*)

I have been a choral singer all my active life, and I love the *Requiem* with all my heart. I have sung the work at least a dozen times, spanning more years than Mozart lived (from a first undergraduate performance at age nineteen to a latest effort at age fifty-five). I do not even care to imagine how much poorer life would be without such music. As with any truly great work of human genius (I have, for example, read Darwin's *Origin of Species* once a decade, each time as an entirely fresh and different book), the *Requiem* never fails to instruct and inspire. As Shakespeare said of Cleopatra: "Age cannot wither her, nor custom stale her infinite variety."

Unpredictable contingency, not lawlike order, rules the pathways of history. A little branch called *Homo sapiens* inhabits this earth by good luck built upon incalculably small probability. Do we not all yearn for the power to tweak those probabilities just a teeny little bit—to replay the tape of history with an apparently inconsequential change that cascades to colossal effect in subsequent times? Suppose, following this greatest of undoable thought experiments, that we alter nothing until 1791, but then let Mozart live to 1830, thus matching Handel's age. Can we even contemplate the added joy, measured in quanta of pleasure for billions of people, provided by another forty symphonies and a dozen operas, perhaps on such sublime texts as *Hamlet, Faust,* or *Lear*? Can we even imagine how differently the history of music, and of human creativity in general, might have run in this ever-so-slightly altered circumstance?

We should, I think, count our blessings instead. Let us not lament an early death at less than half Handelian age. Let us rejoice that smallpox, typhoid, or rheumatic fever (all of which he suffered as a child) did not extinguish Barrington's prodigy before he could grow up to become Mozart. If he had died after *Mitridate* (a teenaged opera of indifferent status), Mozart might only have become a footnote for lamentation. Instead, we received the most sublime swan song ever written—this *Requiem,* fitted with a closing text that might well be read as a prayer of thanks for the sublime gift that Mozart gave to all humanity, and for all time, with his music: *lux aeterna,* eternal light.

13

More Power to Him

IN 1927, WHEN MY FATHER TURNED TWELVE, AL JOLSON inaugurated the era of sound movies with *The Jazz Singer,* Jerome Kern and Oscar Hammerstein's *Show Boat* opened on Broadway, Charles Lindbergh flew the *Spirit of Saint Louis* across the Atlantic nonstop to Paris, the state of Massachusetts executed Sacco and Vanzetti, and Babe Ruth hit 60 home runs in a single season.

Roger Maris bested the Babe with 61 in 1961, the summer of my nineteenth birthday—with teammate Mickey Mantle batting just afterward, and reaching 54 in one of the two greatest home run derbies in baseball history. This summer, Mark McGwire will surely break 61, and may even reach 70* (with Sammy Sosa just behind, or perhaps in front, in the other greatest derby ever). My two sons, both fans in their different ways, will turn twenty-nine and twenty-five.

*I wrote this piece for *The Wall Street Journal* to honor McGwire's sixtieth homer and the certainty of his fracturing Maris's old record of 61. Since nearly every forecast I have ever made has been ludicrously wrong, I do take some pride in the only example I can cite of a personal prediction that, for reasons of pure dumb luck, happened to come up golden. McGwire ended his season with exactly 70 dingers, Sosa with 66.

This magic number, this greatest record in American sports, obsesses us for at least three good reasons. First, baseball has changed no major rule in a century, and we can therefore look and compare, in genuine continuity, across the generations. The seasons of our lives move inexorably forward. As my father saw Ruth, I followed Maris, and my sons watch McGwire. But the game also cycles in glorious sameness, as each winter of our discontent yields to another spring of opening day.

Second, baseball records have clear meaning as personal accomplishments, while marks in most team sports can be judged only as peculiar amalgams. Wilt Chamberlain once scored one hundred points in a single basketball game, but only because his teammates, for that day, elected the odd strategy of feeding him on essentially every play. Home runs are *mano a mano,* batter against pitcher.

Third, and how else can I say this, baseball is just one helluva terrific game, long paramount in American sporting myths and athletic traditions—with the power and definitiveness of a home run as the greatest icon of all. You might argue that Babe Ruth failed to catch the ecumenical spirit when he said, in his famous and moving speech at Yankee Stadium in 1947, as Major League Baseball gathered to honor its dying hero: "The only real game in the world, I think, is baseball. . . . You've got to start from way down . . . when you're six or seven . . . you've got to let it grow up with you." But who would deny the Babe's heartfelt sentiment?

As a veteran and close student of the 1961 Mantle-Maris derby, I thrill to the detailed similarity of McGwire versus Sosa. The two Yankees of 1961 embodied different primal myths about great accomplishments: Mantle, the deserving hero working all his life toward his year of destiny; Maris, the talented journeyman enjoying that one sweet interval in each man's life when everything comes together in some oddly miraculous way. (Maris never hit more than 39 in any other season.) That year, the miracle man won—and more power to him (and shame on his detractors). Fluke or destiny doesn't matter; Roger Maris did the deed.

Mark McGwire is this year's Mantle. No one since Ruth has been so destined, and no one has ever worked harder, and more single-mindedly, to harness and fulfill his gifts of brawn. He is the real item, and this is his year. No one, even Ruth, ever hit more than 50 homers in three successive seasons—as McGwire has now done. (But will anyone ever break Ruth's feat of hitting more than 40 in every year from 1920 to 1932, except for two seasons when injuries caused him to miss more than forty games? Hank Aaron, on the other hand, played as a marvel of consistency over twenty-three seasons. But he never hit more than 47 in a single year, and only once did he hit 40 or more in two successive seasons.) Sammy Sosa is this year's Maris, rising from who-knows-where to challenge the man of destiny. More power to both men.

But we rightly focus on McGwire for the eerie and awesome quality of his particular excellence. Most great records descend in small and even increments from the leader, and no single figure stands leagues ahead of all other mere mortals. The home run record used to follow this conventional pattern: Maris with 61, Ruth with 60, Ruth again with 59, Foxx, Greenberg, and McGwire (last season) with 58, and Wilson and Griffey (also last season) with 56.

But a few champions stand so far above the second-place finisher that we almost have to wonder whether such a leader really belongs within *Homo sapiens.* Consider DiMaggio's fifty-six-game hitting streak in 1941 (regarded by most sports statisticians, myself included, as the most improbable figure in the history of American athletics),* compared with second-place Keeler and Rose, both far away at 44; or Jim Thorpe's lopsided victories in both the pentathlon and decathlon of the 1912 Olympics; or, marking a single man's invention of the art of home run hitting, Babe Ruth's first high figure of 54 in 1920—for with this number he exceeded, all by his lonesome, the sum total for any other entire team in the American League!

McGwire belongs to this most select company of superhuman achievers. He may well hit 70, thus creating the same sweep of empty space that separates DiMaggio and Thorpe from their closest competitors. Moreover, the character of his blasts almost defies belief. A 400-foot home run, while not rare, deserves notice and inspires pride. The vast majority of Major League dingers fall between 300 and 400. Well, only 18 of McGwire's first 60 failed to reach 400 feet, and several have exceeded 500—a figure previously achieved only once every few years among all players combined.

When faced with such an exceptional accomplishment, we long to discover particular reasons. But I think that such a search only denotes a deep fallacy of human thought. No special reason need be sought beyond the good fortune of many effectively random moments grafted upon the guaranteed achievements of the greatest home run hitter in the history of baseball. I don't care if the thin air of Colorado encourages home runs. I don't care if expansion has diluted pitching. I don't care if the ball is livelier or the strike zone smaller. And I deeply don't care if McGwire helps himself to train by taking an over-the-counter

*For the details and documentation of this claim, see chapter 15.

substance regarded as legal by Major League Baseball.* (What nonsense to hold McGwire in any way accountable—simply because we fear that kids may ape him as a role model—for an issue entirely outside his call, and fully in the province of baseball's rule-makers. Let no such hypocrisy dim this greatest moment in our sporting life!)

Mark McGwire has prevailed by creating, in his own person, an ultimate fusion between the two great natural forces of luck and dedicated effort: the gift of an extraordinary body, with the skill of a steadfast dedication to training and study that can only merit the literal meaning of a wonderful word—*enthusiasm,* or "the intake of God."

*Don't get me started on the illogic and hypocrisy of public attitudes to drugs—a real and tragic problem fueled, rather than helped, by our false taxonomies and hyped moralisms that suppress and paralyze effective thought. McGwire (and many other ballplayers) takes androstenedione, now sold at nutrition stores, entirely legally and over the counter (and overtly advertised, not hidden in drawers and available only by request—as druggists sold condoms in my youth). If baseball eventually decides to ban the substance because it may raise testosterone levels, shall we retrospectively denounce McGwire for obeying the law of his time? Do we annul the records of all artists, intellectuals, politicians, and actors who thought that smoking enhanced their performance by calming their nerves?

De Mortuis When Truly *Bonum*

14

Bright Star Among Billions*

AS SAUL DESPISED DAVID FOR RECEIVING TEN THOUsand cheers to his own mere thousand, scientists often stigmatize, for the same reason of simple jealousy, the good work done by colleagues for our common benefit. We live in a philistine nation filled with Goliaths, and we know that science feeds at a public trough. We therefore all give lip service to the need for clear and supportive popular presentation of our work. Why then do we downgrade the professional reputation of colleagues who can convey the power and beauty of science to the hearts and minds of a fascinated, if generally uninformed public?

This narrow-minded error—our own philistinism—arises in part from our general ignorance of the long and honorable literary tradition of popular presentation for science, and our consequent mistake in equating popularization with trivialization, cheapening,

*Originally written as an editorial for *Science,* the leading professional journal of the trade—hence the mode of address to professional researchers, rather than to the general public.

or inaccuracy. Great scientists have always produced the greatest popularizations, without compromising the integrity of subject or author. In the seventeenth century, Galileo wrote both his major books as dialogues in Italian for generally literate readers, not as formal Latin treatises designed only for scholars. In the eighteenth century, the Swiss savant J. J. Scheuchzer produced the beautifully elaborate eight-volume *Physica sacra,* with 750 full-page copperplate engravings illustrating the natural history behind all biblical events. In the nineteenth century, Charles Darwin wrote *The Origin of Species,* the most important and revolutionary of all scientific works, as a book for general readers. (My students often ask me where they can find the technical monograph that served as the basis of Darwin's popular work; I tell them that *The Origin of Species* fulfills both allied, not opposing, functions.)

With the death of Carl Sagan, we have lost both a fine scientist and the greatest popularizer of the twentieth century, if not of all time. In his many books, and especially in his monumental television series *Cosmos*—our century's equivalent of Scheuchzer's *Physica sacra* and the most widely accessed presentation of our subject in all human history—Carl explained the method and content of our discipline to the general public. He also conveyed the excitement of discovery with an uncanny mix of personal enthusiasm and clear presentation unequaled by any predecessor. I mourn his passing primarily because I have lost a dear friend, but I am also sad that many scientists never appreciated his excellence or his importance to all of us, while a few of the best of us (in a shameful incident at the National Academy of Sciences) actively rejected him. (Carl was a remarkably sanguine man, but I know that this incident hurt him deeply.) Too many of us never grasped his legendary service to science.

I would epitomize Carl Sagan's excellence and integrity in three points. First, in an age characterized by the fusion of high and pop culture, Carl moved comfortably across the entire spectrum while never compromising scientific content. He could joke with Johnny Carson, compose a column for *Parade,* and write a science fiction novel while maintaining an active laboratory and publishing technical papers. He had foibles aplenty; don't we all? We joked about his emphatic pronunciation of "billions," and my young son (much to Carl's amusement) called *Cosmos* the "stick-head-up show" because Carl always looked up dreamily into the heavens. But the public watched, loved, and learned. Second, for all his pizzazz and charisma, Carl always spoke for true science against the plethora of irrationalisms that surround us. He conveyed one consistent message: real science is so damned exciting, transforming, and provable; why would anyone prefer the undocumentable nonsense of astrology, alien abductions, and so forth? Third, he bridged the gaps between our various cultures by showing the personal, humanistic, and artistic side of scientific activity. I will never, for example, forget his excellent treatment of Hypatia, a great woman, philosopher, and mathematician, martyred in Alexandria in A.D. 415.

You had a wonderful life, Carl, but far too short. You will, however, always be with us, especially if we as a profession can learn from you how the common touch enriches science while extending an ancient tradition that lies at the heart of Western humanism, and does not represent (when properly done) a journalistic perversion of the "sound bite" age. In the words that John Dryden wrote about another great artist, the composer Henry Purcell, who died even younger in 1695: "He long ere this had tuned the jarring spheres and left no hell below."

15

The Glory of His Time and Ours

IN OUR SAGAS, MOURNING MAY INCLUDE CELEBRATION when the hero dies, not young and unfulfilled on the battlefield, but rich in years and replete with honor. And yet for me, the passing of Joe DiMaggio has evoked a primary feeling of sadness for something precious that cannot be restored—a loss not only of the man, but also of the splendid image that he represented.

I first saw DiMaggio play near the end of his career in 1950, when I was eight and Joe had his last great season, batting .301 with 32 homers and 122 RBIs. He became my hero, my model, and my mentor, all rolled up into one remarkable man. (I longed to be his replacement in center field, but a guy named Mantle came along and beat me out for the job.) DiMaggio remained my primary hero to the day of his death, and through all the vicissitudes of Ms. Monroe, Mr. Coffee, and Mrs. Robinson.

Even with my untutored child's eyes, I could sense something supremely special about DiMaggio's play. I didn't even know the words or their meanings, but I grasped his gracefulness in some visceral way, and I knew that an aura of majesty surrounded all his actions. He played every aspect of baseball with a fluid beauty in minimal motion, a spare elegance that made even his swinging strikeouts look beautiful (an infrequent occurrence in his career; no other leading home run hitter has ever posted more than twice as many lifetime walks as strikeouts or, even more amazingly, nearly as many homers as whiffs—361 dingers versus 369 Ks. Compare this with his two great Yankee long-ball compatriots: 714 homers and 1330 Ks for Ruth, 536 homers and 1710 Ks for Mantle).

His stance, his home run trot, those long flyouts to the cavernous left-center space in Yankee Stadium, his apparently effortless loping run—no hot dog he—to arrive under every catchable fly ball at exactly the right place and time for an "easy" out. If the sports cliché of "poetry in motion" ever held real meaning, DiMaggio must have been the intended prototype.

One cannot extract the essence of DiMaggio's special excellence from the heartless figures of his statistical accomplishments. He did not play long enough to amass leading numbers in any category—only thirteen full seasons from 1936 to 1951, with prime years lost to war, and a fierce pride that led him to retire the moment his skills began to erode.

DiMaggio sacrificed other records to the customs of his time. He hit a career high .381 in 1939, but would probably have finished well over .400 if manager Joe McCarthy hadn't insisted that he play every day in the season's meaningless last few weeks, long after the Yanks had clinched the pennant, while DiMaggio (batting .408 on September 8) then developed such serious sinus problems that he lost sight in one eye, could not visualize in three dimensions, and consequently slipped nearly 30 points in batting average. In those different days, if you could walk, you played.

DiMaggio's one transcendent numerical record—his fifty-six-game hitting streak in 1941—deserves the usual accolade of most remarkable sporting episode of the century, Mark McGwire notwithstanding. Several years ago, I performed a fancy statistical analysis on the data of slumps and streaks, and found that only DiMaggio's shouldn't have happened. All other streaks fall within the expectations for great events that should occur once as a consequence of probabilities, just as an honest coin will come up heads ten times in a row once in a very rare while. But no one should ever have hit in fifty-six straight games. Second place stands at a distant forty-four, a figure reached by Pete Rose and Wee Willie Keeler.

DiMaggio's greatest record therefore embodies pure heart, not the rare expectation of luck. We must also remember that third baseman Ken Keltner robbed DiMaggio of two hits in the fifty-seventh game, and that he then went on to hit safely in sixteen straight games thereafter. DiMaggio also compiled a sixty-one-game hit streak when he played for the San Francisco Seals in the minor Pacific Coast League.

One afternoon in 1950, I sat next to my father near the third base line in Yankee Stadium. DiMaggio fouled a ball in our direction, and my father caught it. We mailed the precious relic to the great man, and sure enough, he sent it back with his signature. That ball remains my proudest possession to this day. Forty years later, during my successful treatment for a supposedly incurable cancer, I received a small square box in the mail from a friend and book publisher in San Francisco, and a golfing partner of DiMaggio. I opened the box and found another ball, signed to me by DiMaggio (at my friend's instigation) and wishing me well in my recovery. What a thrill and privilege—to tie my beginning and middle life together through the good wishes of this great man.

Ted Williams is, appropriately, neither a modest nor a succinct man. When asked recently to compare himself with his rival and contemporary DiMaggio, the greatest batter in history simply replied: "I was a better hitter, he was a better player."

Simon and Garfunkel captured the essence of this great man in their famous lyric about the meaning and loss of true stature: "Where have you gone, Joe DiMaggio? A nation turns its lonely eyes to you."*

He was the glory of a time that we will not see again.

---

*DiMaggio, so wholly possessed of integrity and refinement both on and off the field, was also a very concrete man of few words. In his op-ed obituary for *The New York Times,* Paul Simon tells a wonderful story of his only meeting with DiMaggio and their contretemps over Mrs. Robinson:

> A few years after "Mrs. Robinson" rose to No. 1 on the pop charts, I found myself dining at an Italian restaurant where DiMaggio was seated with a party of friends. I'd heard a rumor that he was upset with the song and had considered a lawsuit, so it was with some trepidation that I walked over and introduced myself as its composer. I needn't have worried: he was perfectly cordial and invited me to sit down, whereupon we immediately fell into conversation about the only subject we had in common.
>
> "What I don't understand," he said, "is why you ask where I've gone. I just did a Mr. Coffee commercial, I'm a spokesman for the Bowery Savings Bank and I haven't gone anywhere."

16

This Was a Man

WHEN MEL ALLEN, "VOICE OF THE YANKEES," DIED LAST week,* I lost the man who ranked second only to my father for sheer volume of attention during my childhood. (My dad, by the way, was a Dodger fan and a Red Barber devotee.) As I considered the surprising depth of my sadness, I realized that I was mourning the extinction of a philosophy as much as the loss of a dear man—and I felt that most of the warm press commentary had missed the essence of Mel Allen's strength. The eulogies focused on his three signature phrases: his invariable opening line, "Hello there, everybody"; his perennial exclamation of astonishment, "How about that"; and his inevitable home run mantra, "It's going . . . going . . . gone."

But I would characterize his immense appeal by two singular statements, one-off comments that I heard in passing moments during a distant childhood. These comments have stayed with me all my life, for integrity in one case, and for antic humor in the other.

*This piece originally appeared in the *New York Times* on June 26, 1996.

One exemplifies the high road, the other an abyss, however charming. The comments could not be more different, but they embody, when taken together, something precious, something fragile, and something sadly lost when institutions become so large that the generic blandness of commercial immensity chokes off both spontaneity and originality. This phenomenon of modern life, by the way, is entirely general and not confined to broadcasting. In my own academic world, textbooks have become longer, duller, and entirely interchangeable for the same reason. Idiosyncratic works cannot sell sufficiently, for curricula have been standardized (partly by the sameness of conventional textbooks)—and originality guarantees oblivion. Authors have become cogs in an expensive venture that includes, among others, the photo researcher, the slide maker, the teacher's guide preparer, and the publicist. The great texts of the past defined fields for generations because they promulgated the distinctive views of brilliant authors—Lyell's geology, or Marshall's economics—but modern writers are faceless servants of a commercial machine that shuns anything unique.

One day in 1952, as Mickey Mantle struggled in center field the year after Joe DiMaggio's retirement, many fans began to boo after Mickey struck out for the second time in a row. In the midst of his play-by-play broadcast, an infuriated Mel Allen leaned out of the press box and shouted at a particularly raucous fan: "Why are you booing him?" The fan shot back: "Because he's not as good as DiMaggio." And Mel Allen busted a gut, delivering a ferocious dressing-down to the fan for his indecency in razzing an enormously talented but unformed twenty-year-old kid just because he could not yet replace the greatest player of the age.

Ballantine beer and White Owl cigars sponsored the Yankees in those years—and Mel never lost an opportunity for additional endorsement. Home runs, for example, became "Ballantine Blasts" or "White Owl Wallops," depending on the sponsor of the inning. When a potential home run passed just to the wrong side of the foul pole, Allen would exclaim, "Why that ball was just foul by a bottle of Ballantine beer." One day Mickey Mantle hit one that seemed destined for success, and Allen began his mantra: "It's going . . . going . . ." And then he stopped short as the ball went foul by no more than an inch or two. An astonished Allen exclaimed: "Why, I've never seen one miss by so little. That ball was foul by no more than a bottle of Bal—" And then he paused in mid phrase, thought for a fraction of a moment, and exclaimed: "No, that ball was foul by the ash on a White Owl cigar!"

A man of grace and integrity; a shameless huckster of charming originality. But above all, a man who could only be his wonderful cornball self—Mel Allen, the singular, inimitable, human Voice of the Yankees. So take my two stories, titrate them to the optimal distinctness of lost individuality, and let us celebrate Shakespeare's judgment in *Julius Caesar:* "The elements so mix'd in him that Nature might stand up and say to the world, 'This was a man!'"

V

Science in Society

17

A Tale of Two Work Sites

CHRISTOPHER WREN, THE LEADING ARCHITECT OF LONDON'S reconstruction after the great fire of 1666, lies buried beneath the floor of his most famous building, St. Paul's cathedral. No elaborate sarcophagus adorns the site. Instead, we find only the famous epitaph written by his son and now inscribed into the floor: *"si monumentum requiris, circumspice"*—if you are searching for his monument, look around. A tad grandiose perhaps, but I have never read a finer testimony to the central importance—one might even say sacredness—of actual places, rather than replicas, symbols, or other forms of vicarious resemblance.

An odd coincidence of professional life turned my thoughts to this most celebrated epitaph when, for the second time, I received an office in a spot laden with history, a place still redolent of ghosts of past events both central to our common culture and especially meaningful for my own life and choices.

In 1971, I spent an academic term as a visiting researcher at

Oxford University. I received a cranny of office space on the upper floor of the University Museum. As I set up my books, fossil snails, and microscope, I noticed a metal plaque affixed to the wall, informing me that this reconfigured space of shelves and cubicles had been, originally, the site of the most famous public confrontation in the early history of Darwinism. On this very spot, in 1860, just a few months after Darwin published *The Origin of Species,* T. H. Huxley had drawn his rhetorical sword, and soundly skewered the slick but superficial champion of creationism, Bishop "Soapy Sam" Wilberforce.

(As with most legends, the official version ranks as mere cardboard before a much more complicated and multifaceted truth. Wilberforce and Huxley did put on a splendid, and largely spontaneous, show—but no clear victor emerged from the scuffle, and Joseph Hooker, Darwin's other champion, made a much more effective reply to the bishop, however forgotten by history. See my essay on this debate, entitled "Knight Takes Bishop?" and published in an earlier volume of this series, *Bully for Brontosaurus.*)

I can't claim that the lingering presence of these Victorian giants increased my resolve or improved my work, but I loved the sense of continuity vouchsafed to me by this happy circumstance. I even treasured the etymology—for *circumstance* means "standing around" (as Wren's *circumspice* means "looking around"), and here I stood, perhaps in the very spot where Huxley had said, at least according to legend, that he preferred an honest ape for an ancestor to a bishop who would distort a known truth for rhetorical advantage.

Not so long ago, I received a part-time appointment as visiting research professor of biology at New York University. I was given an office on the tenth floor of the Brown building on Washington Place, a nondescript early-twentieth-century structure now filled with laboratories and other academic spaces. As the dean took me on a casual tour of my new digs, he made a passing remark, intended as little more than "tour-guide patter," but producing an electric effect upon his new tenant. Did I know, he asked, that this building had been the site of the infamous Triangle Shirtwaist fire of 1911, and that my office occupied a corner location on one of the affected floors—in fact, as I later discovered, right near the escape route used by many workers to safety on the roof above. The dean also told me that, each year on the March 25 anniversary of the fire, the International Ladies' Garment Workers Union still holds a ceremony at the site and lays wreaths to memorialize the 146 workers killed in the blaze.

If the debate between Huxley and Wilberforce defines a primary legend of my chosen profession, the Triangle Shirtwaist fire occupies an even more central place in my larger view of life. I grew up in a family of Jewish immigrant garment workers, and this holocaust (in the literal meaning of a thorough sacrifice by burning) had set their views and helped to define their futures.

The shirtwaist—a collared blouse designed on the model of a man's shirt and worn above a separate skirt—had become the fashionable symbol of more independent women. The Triangle Shirtwaist Company, New York City's largest manufacturer of shirtwaists, occupied three floors (eighth through tenth) of the Asch Building (later bought by New York University and rechristened Brown, partly to blot out the infamy of association with the fire). The company employed some five hundred workers, nearly all young women who had recently arrived either as Jewish immigrants from eastern Europe or as Catholics from Italy. Exits from the building, in addition to elevators, included only two small stairways and one absurdly inadequate fire escape. But the owners had violated no codes, both because general standards of regulation were then so weak, and because the structure was supposedly fireproof—as the framework proved to be (for the building, with my office, still stands), though inflammable walls and ceilings could not prevent an internal blaze on floors crammed full of garments and cuttings. The Triangle company was, in fact, a deathtrap—for fire hoses of the day could not pump above the sixth floor, while nets and blankets could not sustain the force of a human body jumping from greater heights.

The fire broke out at quitting time. Most workers managed to escape by the elevators, down one staircase (we shall come to the other staircase later), or by running up to the roof. But the flames trapped 146 employees, nearly all young women. About fifty workers met a hideous, if dramatic, end by jumping in terror from the ninth-floor windows, as a wall of fire advanced from behind. Firemen and bystanders begged them not to jump, and then tried to hold improvised nets of sheets and blankets. But these professionals and good Samaritans could not hold the nets against the force of fall, and many bodies plunged right through the flimsy fabrics onto the pavement below, or even through the "hollow sidewalks" made of opaque glass circles designed to transmit daylight to basements below, and still a major (and attractive) feature of my SoHo neighborhood. (These sidewalks carry prominent signs warning delivery trucks not to back in.) Not a single jumper survived, and the memory of these forced leaps to death remains the most searing image of America's prototypical sweatshop tragedy.

All defining events of history develop simplified legends as official versions—primarily, I suppose, because we commandeer such events for shorthand moral instruction, and the complex messiness of actual truth always blurs the clarity of a pithy epigram. Thus, Huxley, representing the righteousness of scientific objectivity, must slay the dragon of ancient and unthinking dogma. The equally oversimplified legend of the Triangle fire holds that workers became trapped because management had locked all the exit doors to prevent pilfering, unscheduled breaks, or access to union organizers—leaving only the fire escape as a mode of exit. All five of my guidebooks to New York architecture tell this "official" version. My favorite book, for example, states: "Although the building was equipped with fire exits, the terrified workers discovered to their horror that the ninth-floor doors had been locked by supervisors. A single fire-escape was wholly inadequate for the crush of panic-stricken employees."

These traditional (indeed, virtually "official") legends may exaggerate for moral punch, but such interpretations emerge from a factual basis of greater ambiguity—and this reality, as we shall see in the Triangle case, often embodies a deeper and more important lesson. Huxley did argue with Wilberforce, after all, even if he secured no decisive victory, and Huxley did represent the side of the angels—the true angels of light and justice. And although many Triangle workers escaped by elevators and one staircase, another staircase (that might have saved nearly everyone else) was almost surely locked.

If Wilberforce and his minions had won, I might be a laborer, a linguist, or a lawyer today. But the Triangle fire might have blotted me out entirely. My grandmother arrived in America in 1910. On that fatal March day in 1911, she was working as a sixteen-year-old seamstress in a sweatshop—but, thank God, not for the Triangle Shirtwaist Company. My grandfather, at the same moment, was cutting cloth in yet another nearby factory.

These two utterly disparate stories—half a century and an ocean apart, and with maximal contrast between an industrial tragedy and an academic debate—might seem to embody the most unrelatable of items: the apples and oranges, or chalk and cheese (the British version), of our mottoes. Yet I feel that an intimate bond ties these two stories together in illustrating opposite poles of a central issue in the history of evolutionary theory: the application of Darwinian thought to the life and times of our own troubled species. I claim nothing beyond personal meaning—and certainly no rationale for boring anyone else—in the accidental location of my two offices in such sacred spots of history. But the emotion of a personal prod often dislodges a general theme well worth sharing.

The application of evolutionary theory to *Homo sapiens* has always troubled Western culture deeply—not for any reason that might be called scientific (for humans are biological objects, and must therefore take their place with all other living creatures on the genealogical tree of life), but only as a consequence of ancient prejudices about human distinctiveness and unbridgeable superiority. Even Darwin tiptoed lightly across this subject when he wrote *The Origin of Species* in 1859 (though he plunged in later, in 1871, by publishing *The Descent of Man*). The first edition of the *Origin* says little about *Homo sapiens* beyond a cryptic promise that "light will be thrown on the origin of man and his history." (Darwin became a bit bolder in later editions and ventured the following emendation: "Much light will be thrown . . .")

Troubling issues of this sort often find their unsurprising resolution in a bit of wisdom that has permeated our traditions from such sublime sources as Aristotle's *aurea mediocritas* (or golden mean) to the vernacular sensibility of Goldilocks's decisions to split the difference between two extremes, and find a solution "just right" in the middle. Similarly, one can ask either too little or too much of Darwinism in trying to understand "the origin of man and his history." As usual, a proper solution lies in the intermediary position of "a great deal, but not everything." Soapy Sam Wilberforce and the Triangle Shirtwaist fire gain their odd but sensible conjunction as illustrations of the two extremes that must be avoided—for Wilberforce denied evolution altogether and absolutely, while the major social theory that hindered industrial reform (and permitted conditions that led to such disasters as the Triangle Shirtwaist fire) followed the most overextended application of biological evolution to patterns of human history—the theory of "Social Darwinism." By understanding the fallacies of Wilberforce's denial and social Darwinism's uncritical and total embrace, we may find the proper balance between.

They didn't call him Soapy Sam for nothing. The orotund bishop of Oxford saved his finest invective for Darwin's attempt to apply his heresies to human origins. In his review of *The Origin of Species* (published in the *Quarterly Review*, England's leading literary journal, in 1860), Wilberforce complained above all: "First, then, he not obscurely declares that he applies his scheme of the action of the principle of natural selection to Man himself, as well as to the animals around him." Wilberforce then uncorked a passionate argument for a human uniqueness that could only have been divinely ordained:

> Man's derived supremacy over the earth; man's power of articulate speech; man's gift of reason; man's free-will and responsibility; man's fall and man's redemption; the incarnation of the Eternal Son; the indwelling of the Eternal Spirit,—all are equally and

> utterly irreconcilable with the degrading notion of the brute origin of him who was created in the image of God, and redeemed by the Eternal Son.

But the tide of history quickly engulfed the good bishop. When Wilberforce died in 1873, from a head injury after a fall from his horse, Huxley acerbically remarked that, for once, the bishop's brains had come into contact with reality—and the result had been fatal. Darwinism became the reigning intellectual novelty of the late nineteenth century. The potential domain of natural selection, Darwin's chief explanatory principle, seemed nearly endless to his devotees (though not, interestingly, to the master himself, as Darwin remained cautious about extensions beyond the realm of biological evolution). If a "struggle for existence" regulated the evolution of organisms, wouldn't a similar principle also explain the history of just about anything—from the cosmology of the universe, to the languages, economics, technologies, and cultural histories of human groups?

Even the greatest of truths can be overextended by zealous and uncritical acolytes. Natural selection may be one of the most powerful ideas ever developed in science, but only certain kinds of systems can be regulated by such a process, and Darwin's principle cannot explain all natural sequences that develop historically. For example, we may talk about the "evolution" of a star through a predictable series of phases over many billion years from birth to explosion, but natural selection—a process driven by the differential survival and reproductive success of some individuals in a variable population—cannot be the cause of stellar development. We must look, instead, to the inherent physics and chemistry of light elements in such large masses.

Similarly, although Darwinism surely explains many universal features of human form and behavior, we cannot invoke natural selection as the controlling cause of our cultural changes since the dawn of agriculture—if only because such a limited time of some ten thousand years provides so little scope for any general biological evolution at all. Moreover, and most importantly, human cultural change operates in a manner that precludes a controlling role for natural selection. To mention the two most obvious differences: first, biological evolution proceeds by continuous division of species into independent lineages that must remain forever separated on the branching tree of life. Human cultural change works by the opposite process of borrowing and amalgamation. One good look at another culture's wheel or alphabet may alter the course of a civilization forever. If we wish to identify a biological analog for cultural change, I suspect that infection will work much better than evolution.

Second, human cultural change runs by the powerful mechanism of Lamarckian inheritance of acquired characters. Anything useful (or alas, destructive) that our generation invents can be passed directly to our offspring by direct education. Change in this rapid Lamarckian mode easily overwhelms the much slower process of Darwinian natural selection, which requires a Mendelian form of inheritance based on small-scale and undirected variation that can then be sifted and sorted through a struggle for existence. Genetic variation is Mendelian, so Darwinism rules biological evolution. But cultural variation is largely Lamarckian, and natural selection cannot determine the recent history of our technological societies.

Nonetheless, the first blush of high Victorian enthusiasm for Darwinism inspired a rush of attempted extensions to other fields, at least by analogy. Some efforts proved fruitful, including the decision of James Murray, editor of *The Oxford English Dictionary* (first volume published in 1884, but under way for twenty years before then), to work strictly by historical principles and to treat the changing definitions of words not by current preferences in use (as in a truly normative dictionary), but by the chronology and branching evolution of recorded meanings (making the text more an encyclopedia about the history of words than a true dictionary).

But other extensions proved both invalid in theory, and also (or so most of us would judge by modern moral sensibilities) harmful, if not tragic, in application. As the chief offender in this category, we must cite a highly influential theory that acquired the inappropriate name of "Social Darwinism." (As many historians have noted, this theory should really be called "social Spencerism," since Herbert Spencer, chief Victorian pundit of nearly everything, laid out all the basic postulates in his *Social Statics* of 1850, nearly a decade before Darwin published *The Origin of Species.* Darwinism did add the mechanism of natural selection as a harsher version of the struggle for existence, long recognized by Spencer. Moreover, Darwin himself maintained a highly ambivalent relationship to this movement that borrowed his name. He felt the pride of any creator toward useful extensions of his theory—and he did hope for an evolutionary account of human origins and historical patterns. But he also understood only too well why the mechanism of natural selection applied poorly to the causes of social change in humans.)

Social Darwinism often serves as a blanket term for any genetic or biological claim made about the inevitability (or at least the "naturalness") of social inequalities among classes and sexes, or military conquests of one group by another. But such a broad definition distorts the history of this important subject—although pseudo-Darwinian arguments have long been advanced, prominently and forcefully, to cover all these sins. Classical Social Darwinism operated as a more specific theory about the nature and origin of social classes in the modern industrial world. The *Encyclopaedia Britannica* article on this subject correctly emphasizes this restriction by first citing the broadest range of potential meaning, and then properly narrowing the scope of actual usage:

> *Social Darwinism:* the theory that persons, groups, and races are subject to the same laws of natural selection as Charles Darwin had perceived in plants and animals in nature. . . . The theory was used to support laissez-faire capitalism and political conservatism. Class stratification was justified on the basis of "natural" inequalities among individuals, for the control of property was said to be a correlate of superior and inherent moral attributes such as industriousness, temperance, and frugality. Attempts to reform society through state intervention or other means would, therefore, interfere with natural processes; unrestricted competition and defense of the status quo were in accord with biological selection. The poor were the "unfit" and should not be aided; in the struggle for existence, wealth was a sign of success.

Spencer believed that we must permit and welcome such harshness to unleash the progressive development that all "evolutionary" systems undergo if allowed to follow their natural course in an unimpeded manner. As a central principle of his system, Spencer believed that progress—defined by him as movement from a simple undifferentiated homogeneity, as in a bacterium or a "primitive" human society without social classes, to complex and structured heterogeneity, as in "advanced" organisms or industrial societies—did not arise as an inevitable property of matter in motion, but only through interaction between evolving systems and their environments. These interactions must therefore not be obstructed.

The relationship of Spencer's general vision to Darwin's particular theory has often been misconstrued or overemphasized. As stated above, Spencer had published the outline (and most of the details) of his system nearly ten years before Darwin presented his evolutionary theory. Spencer certainly did welcome the principle of natural selection as an even more ruthless and efficient mechanism for driving evolution forward. (Ironically, the word *evolution,* as a description for the genealogical history of life, entered our language through Spencer's urgings, not from Darwin. Spencer favored the term for its vernacular English meaning of "progress," in the original Latin sense of *evolutio,* or "unfolding." At first, Darwin resisted the term—he originally called his process "descent with modification"—because his theory included no mechanism or rationale for general progress in the history of life. But Spencer prevailed, largely because no society has ever been more committed to progress as a central notion or goal than Victorian Britain at the height of its colonial and industrial expansion.)

Spencer certainly used Darwin's mechanism of natural selection to buttress his system. Few people recognize the following historical irony: Spencer, not Darwin, coined the term "survival of the fittest," now our conventional catchphrase for Darwin's mechanism. Darwin himself paid proper tribute in a statement added to later editions of *The Origin of Species:* "I have called this principle, by which each slight variation, if useful, is preserved, by the term Natural Selection. . . . But the expression often used by Mr. Herbert Spencer of the Survival of the Fittest is more accurate, and is sometimes equally convenient."

As a mechanism for driving his universal "evolution" (of stars, species, languages, economics, technologies, and nearly anything else) toward progress, Spencer preferred the direct and mechanistic "root, hog, or die" of natural selection (as William Graham Sumner, the leading American social Darwinian, epitomized the process), to the vaguer and largely Lamarckian drive toward organic self-improvement that Spencer had originally favored as a primary cause. (In this colorful image, Sumner cited a quintessential American metaphor of self-sufficiency that my dictionary of catchphrases traces to a speech by Davy Crockett in 1834.) In a post-Darwinian edition of his *Social Statics,* Spencer wrote:

> The lapse of a third of a century since these passages were published, has brought me no reason for retreating from the position taken up in them. Contrariwise, it has brought a vast amount of evidence strengthening that position. The beneficial results of the survival of the fittest, prove to be immeasurably greater than [I formerly recognized]. The process of "natural selection," as Mr. Darwin called it . . . has shown to be a chief cause . . . of that evolution through which all living things, beginning with the lower, and diverging and re-diverging as they evolved, have reached their present degrees of organization and adaptation to their modes of life.

But putting aside the question of Darwin's particular influence, the more important, underlying point remains firm: the theory of Social Darwinism (or social Spencerism) rests upon a set of analogies between the causes of change and stability in biological and social systems—and on the supposedly direct applicability of these biological principles to the social realm. In his founding document, the *Social Statics* of 1850, Spencer rests his case upon two elaborate analogies to biological systems.

1. The struggle for existence as purification in biology and society. Darwin recognized the "struggle for existence" as metaphorical shorthand for any strategy that promotes increased reproductive success, whether by outright battle, cooperation, or just simple prowess in copulation under the old principle of "early and often." But many contemporaries, including Spencer, read "survival of the fittest" only as overt struggle to the death—what T. H. Huxley later dismissed as the "gladiatorial" school, or the incarnation of Hobbes's *bellum omnium contra omnes* (the war of all against all). Spencer presented this stark and limited view of nature in his *Social Statics:*

> Pervading all Nature we may see at work a stern discipline which is a little cruel that it may be very kind. That state of universal warfare maintained throughout the lower creation, to the great perplexity of many worthy people, is at bottom the most merciful provision which the circumstances admit of. . . . Note that carnivorous enemies, not only remove from herbivorous herds individuals past their prime, but also weed out the sickly, the malformed, and the least fleet or powerful. By the aid of which purifying process . . . all vitiation of the race through the multiplication of its inferior samples is prevented; and the maintenance of a constitution completely adapted to surrounding conditions, and therefore most productive of happiness, is ensured.

Spencer then compounds this error by applying the same argument to human social history, without ever questioning the validity of such analogical transfer. Railing against all governmental programs for social amelioration—Spencer opposed state-supported education, postal services, regulation of housing conditions, and even public construction of sanitary systems—he castigated such efforts as born of good intentions but doomed to dire consequences by enhancing the survival of social dregs who should be allowed to perish for the good of all. (Spencer insisted, however, that he did not oppose private charity, primarily for the salutary effect of such giving upon the moral development of donors. Does this discourse remind you of arguments now advanced as reformatory and spanking-new by our "modern" ultraconservatives? Shall we not profit from Santayana's famous dictum that those ignorant of history must be condemned to repeat it?) In his chapter on poor laws (which he, of course, opposed), Spencer wrote in the *Social Statics:*

> We must call those spurious philanthropists who, to prevent present misery, would entail greater misery on future generations. That rigorous necessity which, when allowed to operate, becomes so sharp a spur to the lazy and so strong a bridle to the random, these paupers' friends would repeal, because of the wailings it here and there produces. Blind to the fact that under the natural order of things society is constantly excreting its unhealthy, imbecile, slow, vacillating, faithless members, these unthinking, though well-meaning, men advocate an interference which not only stops the purifying process, but even increases the vitiation—absolutely encouraging the multiplication of the reckless and incompetent by offering them an unfailing provision. . . . Thus, in their eagerness to prevent the salutary sufferings that surround us, these sigh-wise and groan-foolish people bequeath to posterity a continually increasing curse.

2. The stable body and the stable society. In the universal and progressive "evolution" of all systems, organization becomes increasingly more complex by division of labor among the growing number of differentiating parts. All parts must "know their place" and play their appointed role, lest the entire system collapse. A primitive hydra, constructed of simple "all purpose" modules, can regrow any lost part, but nature gives a man only one head, and one chance. Spencer recognized the basic inconsistency in validating social stability by analogy to the integrated needs of a single organic body—for he recognized the contrary rationales of the two systems: the parts of a body serve the totality, but the social totality (the state) supposedly exists only to serve the parts (individual people). But Spencer never allowed himself to be fazed by logical or empirical difficulties when pursuing such a lovely generality. (Huxley was speaking about Spencer's penchant for building grandiose systems when he made his famous remark about "a beautiful theory, killed by a nasty, ugly little fact.") So Spencer barged right through the numerous absurdities of such a comparison, and even claimed that he had found a virtue in the differences. In his famous 1860 article, "The Social Organism," Spencer

described the comparison between a human body and a human society: "Such, then, are the points of analogy and the points of difference. May we not say that the points of difference serve but to bring into clearer light the points of analogy."

Spencer's article then lists the supposed points of valid comparison, including such far-fetched analogies as the historical origin of a middle class to the development, in complex animals, of the mesoderm, or third body layer between the original ectoderm and endoderm; the likening of the ectoderm itself to the upper classes, for sensory organs that direct an animal arise in ectoderm, while organs of production, for such activities as digesting food, emerge from the lower layer, or endoderm; the comparison of blood and money; the parallel courses of nerve and blood vessels in higher animals with the side-by-side construction of railways and telegraph wires; and finally, in a comparison that even Spencer regarded as forced, the likening of a primitive all-powerful monarchy with a simple brain, and an advanced parliamentary system with a complex brain composed of several lobes. Spencer wrote: "Strange as this assertion will be thought, our Houses of Parliament discharge in the social economy, functions that are in sundry respects comparable to those discharged by the cerebral masses in a vertebrate animal."

Spencer surely forced his analogies, but his social intent could not have been more clear: a stable society requires that all roles be filled and well executed—and government must not interfere with a natural process of sorting out and allocation of appropriate rewards. A humble worker must toil, and may remain indigent forever, but the industrious poor, as an organ of the social body, must always be with us:

> Let the factory hands be put on short time, and immediately the colonial produce markets of London and Liverpool are depressed. The shopkeeper is busy or otherwise, according to the amount of the wheat crop. And a potato-blight may ruin dealers in consols. . . . This union of many men into one community—this increasing mutual dependence of units which were originally independent—this gradual segregation of citizens into separate bodies with reciprocally-subservient functions—this formation of a whole consisting of unlike parts—this growth of an organism, of which one portion cannot be injured without the rest feeling it—may all be generalized under the law of individuation.

Social Darwinism grew into a major movement, with political, academic, and journalistic advocates for a wide array of particular causes. But as historian Richard Hofstadter stated in the most famous book ever written on this subject—*Social Darwinism in American Thought,* first published in 1944, in press ever since, and still full of insight despite some inevitable archaisms—the primary impact of this doctrine lay in its buttressing of conservative political philosophies, particularly through the central (and highly effective) argument against state support of social services and governmental regulation of industry and housing:

> One might, like William Graham Sumner, take a pessimistic view of the import of Darwinism, and conclude that Darwinism could serve only to cause men to face up to the inherent hardship of the battle of life; or one might, like Herbert Spencer, promise that, whatever the immediate hardships for a large portion of mankind, evolution meant progress and thus assured that the whole process of life was tending toward some very remote but altogether glorious consummation. But in either case the conclusions to which Darwinism was at first put were conservative conclusions. They suggested that all attempts to reform social processes were efforts to remedy the irremediable, that they interfered with the wisdom of nature, that they could lead only to degeneration.

The industrial magnates of America's gilded age ("robber barons," in a terminology favored by many people) adored and promoted this argument against regulation, evidently for self-serving reasons, and however frequently they mixed their lines about nature's cruel inevitability with standard Christian piety. John D. Rockefeller stated in a Sunday school address:

> The growth of a large business is merely a survival of the fittest. . . . The American Beauty rose can be produced in the splendor and fragrance which bring cheer to its beholder only by sacrificing the early buds which grow up around it. This is not an evil tendency in business. It is merely the working-out of a law of nature and a law of God.

And Andrew Carnegie, who had been sorely distressed by the apparent failure of Christian values, found his solution in Spencer's writings, and then sought out the English philosopher for friendship and substantial favors. Carnegie wrote about his discovery of Spencer's work: "I remember that light came as in a flood and all was clear. Not only had I got rid of theology and the supernatural, but I had found the truth of evolution. 'All is well since all grows better' became my motto, and true source of comfort." Carnegie's philanthropy, primarily to libraries and universities, ranks as one of the great charitable acts of American history, but we should not forget his ruthlessness and resistance to reforms for his own workers (particularly his violent breakup of the Homestead strike of 1892) in building his empire of steel—a harshness that he defended with the usual Spencerian line that any state regulation must derail an inexorable natural process eventually leading to progress for all. In his most famous essay (entitled "Wealth," and published in *North American Review* for 1889), Carnegie stated:

> While the law may be sometimes hard for the individual, it is best for the race, because it insures the survival of the fittest in every department. We accept and welcome, therefore, as conditions to which we must accommodate ourselves, great inequality of environment, the concentration of wealth, business, industrial and commercial, in the hands of a few, and the law of competition between these, as being not only beneficial, but essential for the future progress of the race.

I don't want to advocate a foolishly grandiose view about the social and political influence of academic arguments—and I also wish to avoid the common fallacy of inferring a causal connection from a correlation. Of course I do not believe that the claims of Social Darwinism directly caused the ills of unrestrained industrial capitalism and the suppression of workers' rights. I know that most of these Spencerian lines functioned as mere window dressing for social forces well in place, and largely unmovable by any academic argument.

On the other hand, academic arguments should not be regarded as entirely impotent either—for why else would people in power invoke such claims so forcefully? The general thrust of social change unfolded in its own complex manner without much impact from purely intellectual rationales, but many particular issues—especially the actual rates and styles of changes that would have eventually occurred in any case—could be substantially affected by academic discourse. Millions of people suffered when a given reform experienced years of legislative delay, and then became vitiated in legal battles and compromises. The Social Darwinian argument of the superrich and the highly conservative did stem, weaken, and slow the tides of amelioration, particularly for workers' rights.

Most historians would agree that the single most effective doctrine of Social Darwinism lay in Spencer's own centerpiece—the argument against state-enforced standards for industry, education, medicine, housing, public sanitation, and so on. Few Americans, even the robber barons, would go so far, but Spencerian dogma did become a powerful bludgeon against the regulation of industry to ensure better working conditions for laborers. On this particular point—the central recommendation of Spencer's system from the beginning—we may argue for a substantial effect of academic writing upon the actual path of history.

Armed with this perspective, we may return to the Triangle Shirtwaist fire, the deaths of 146 young workers, and the palpable influence of a doctrine that applied too much of the wrong version of Darwinism to human history. The battle for increased safety of workplaces, and healthier environments for workers, had been waged with intensity for several decades. The trade union movement put substantial priority upon these issues, and management had often reacted with intransigence, or even violence, citing their Spencerian rationale for the perpetuation of apparent cruelty. Government regulation of industry had become a major struggle of American political life—and the cause of benevolent state oversight had advanced from the Sherman Anti-Trust Act of 1890 to the numerous and crusading reforms of Theodore Roosevelt's presidency (1901–9). When the Triangle fire broke out in 1911, regulations for the health and safety of workers were so weak, and so unenforceable by tiny and underpaid staffs, that the company's managers—cynically and technically "up to code" in their firetrap building—could pretty much impose whatever the weak and nascent labor union movement couldn't prevent.

If the standard legend were true—and the Triangle workers died because all the doors had been locked by cruel owners—then this heart-wrenching story might convey no moral beyond the personal guilt of management. But the loss of 146 lives occurred for much more complicated reasons, all united by the pathetic weakness of legal regulations for the health and safety of workers. And I do not doubt that the central thrust of Social Darwinism—the argument that governmental regulation can only forestall a necessary and natural process—exerted a major impact in slowing the passage of laws that almost everyone today, even our archconservatives, regard as beneficial and humane. I accept that these regulations would eventually have been instituted even if Spencer had never been born—but life or death for the Triangle workers rode upon the "detail" that forces of pure laissez-faire, buttressed by their Spencerian centerpiece, managed to delay some implementations to the 1920s, rather than acceding to the just demands of unions and social reformers in 1910.

One of the two Triangle stairways almost surely had been locked on that fateful day—although lawyers for company owners won acquittal of their clients on this issue, largely by using legal legerdemain to confuse, intimidate, and draw inconsistencies from young witnesses with poor command of English. Two years earlier, an important strike had begun at the Triangle company, and had spread to shirtwaist manufacturers throughout the city. The union won in most factories but not, ironically, at Triangle—where management held out, and compelled the return of workers without anything gained. Tensions remained high at Triangle in 1911, and management had become particularly suspicious, even paranoid, about thefts. Therefore, at quitting time (when the fire erupted, and against weakly enforced laws for maintaining multiple active exits), managers had locked one of the doors to force all the women to exit by the Greene Street stairwell, where a supervisor could inspect every handbag to guard against thefts of shirtwaists.

But if the bosses broke a weak and unenforceable law in this instance, all other causes of death can be traced to managerial compliance with absurdly inadequate standards, largely kept so weak by active political resistance to legal regulation of work sites, buttressed by the argument of Social Darwinism. Fire hoses could not pump above the sixth floor, but no law prevented the massing of workers into crowded floors above. No statute required fire drills or other forms of training for safety. In other cases, weak regulations were risibly inadequate, easy to flaunt, and basically unenforced in any case. For example, by law, each worker required 250 cubic feet of air space—a good rule to prevent crowding. But companies had managed to circumvent the intent of this law, and maintain their traditional (and dangerous) density of workers, by moving into large loft buildings with high ceilings and substantial irrelevant space that could be included in calculating the 250-cubic-foot minimum.

When the Asch Building opened in 1900, an inspector for the Buildings Department informed the architect that a third staircase should be provided. But the architect sought and received a variance, arguing that the single fire escape could count as the missing staircase required by law for structures with more than ten thousand square feet per floor. Moreover, the single fire escape—which buckled and fell during the fire, as a result of poor maintenance and the weight of too many workers trying to escape—led only to a glass skylight in a closed courtyard. The building inspector had also complained about this arrangement, and the architect had promised to make the necessary alterations.

But no changes had been made, and the falling fire escape plunged right through the skylight, greatly increasing the death toll.

Two final quotations highlight the case for inadequate legal protection as a primary cause for the unconscionable death toll in the Triangle Shirtwaist fire (Leon Stein's excellent book, *The Triangle Fire,* J. B. Lippincott Company, 1962, served as my chief source for information about this event). Rose Safran, a survivor of the fire and supporter of the 1909 strike, said: "If the union had won we would have been safe. Two of our demands were for adequate fire escapes and for open doors from the factories to the street. But the bosses defeated us and we didn't get the open doors or the better fire escapes. So our friends are dead." A building inspector who had actually written to the Triangle management just a few months before, asking for an appointment to discuss the initiation of fire drills, commented after the blaze: "There are only two or three factories in the city where fire drills are in use. In some of them where I have installed the system myself, the owners have discontinued it. The neglect of factory owners in the matter of safety of their employees is absolutely criminal. One man whom I advised to install a fire drill replied to me: 'Let 'em burn. They're a lot of cattle, anyway.'"

The Triangle fire galvanized the workers' reform movement as never before. An empowered force, now irresistible, of labor organizers, social reformers, and liberal legislators pressed for stronger regulation under the theme of "never again." Hundreds of laws passed as a direct result of this belated agitation. But nothing could wash the blood of 146 workers from the sidewalks of New York.

This tale of two work sites—of a desk situated where Huxley debated Wilberforce, and an office built on a floor that burned during the Triangle Shirtwaist fire—has no end, for the story illustrates a theme of human intellectual life that must always be with us, however imbued with an obvious and uncontroversial solution. Extremes must usually be regarded as untenable, even dangerous places on complex and subtle continua. For the application of Darwinian theory to human history, Wilberforce's "none" marks an error of equal magnitude with the "all" of an extreme Social Darwinism. In a larger sense, the evolution of a species like *Homo sapiens* should fill us with notions of glory for our odd mental uniqueness, and of deep humility for our status as a tiny and accidental twig on such a sturdy and luxuriantly branching tree of life. Glory *and* humility! Since we can't abandon either feeling for a unitary stance in the middle, we had best make sure that both attitudes *always* walk together, hand in hand, and secure in the wisdom of Ruth's promise to Naomi: "Whither thou goest, I will go; and where thou lodgest, I will lodge."

18

The Internal Brand of the Scarlet W

AS A SETTING FOR AN INITIAL WELCOME TO A NEW home, the international arrivals hall at Kennedy airport pales before the spaciousness, the open air, and the symbol of fellowship in New York's harbor. But the plaque that greets airborne immigrants of our time shares one feature with the great lady who graced the arrival of so many seaborne ancestors, including all my grandparents in their childhood. The plaque on Kennedy's wall and the pedestal of the Statue of Liberty bear the same inscription: Emma Lazarus's poem "The New Colossus"—but with one crucial difference. The Kennedy version reads:

> *Give me your tired, your poor,*
> *Your huddled masses yearning to breathe free . . .*
> *Send these, the homeless, tempest-tossed to me:*
> *I lift my lamp beside the golden door.*

One might be excused for supposing that the elision represents a large and necessary omission to fit the essence of a longer poem onto a smallish plaque. But only one line, easily accommodated, has been cut—and for a reason that can only reflect thoughtless (as opposed to merely ugly) censorship, therefore inviting a double indictment on independent charges of stupidity *and* cowardice. (As a member of the last public school generation trained by forced memorization of a holy historical canon, including the Gettysburg Address, the preamble to the Constitution, Mr. Emerson on the rude bridge that arched the flood, and Ms. Lazarus on the big lady with the lamp, I caught the deletion right away, and got sufficiently annoyed to write a *New York Times* op-ed piece a couple of years ago. Obviously, I am still seething, but at least I now have the perverse pleasure of using the story for my own benefit to introduce this essay.) I therefore restore the missing line (along with Emma Lazarus's rhyming scheme and syntax):

The wretched refuse of your teeming shore

Evidently, the transient wind of political correctness precludes such a phrase as "wretched refuse," lest any visitor read the line too literally or personally. Did the authorities at our Port Authority ever learn about metaphor, and its prominence in poetry? Did they ever consider that Ms. Lazarus might be describing the disdain of a foreign elite toward immigrants whom we would welcome, nurture, and value?

This story embodies a double irony that prompted my retelling. We hide Emma Lazarus's line today because we misread her true intention, and because contemporary culture has so confused (and often even equated) inappropriate words with ugly deeds. But the authorities of an earlier generation invoked the false and literal meaning—the identification of most immigrants as wretched refuse—to accomplish a deletion of persons rather than words. The supposed genetic inferiority of most refugees (an innate wretchedness that American opportunity could never overcome) became an effective rallying cry for a movement that did succeed in imposing strong restrictions upon immigration, beginning in the 1920s. These laws, strictly enforced despite pleas for timely exception, immured thousands of Europeans who sought asylum because Hitler's racial laws had marked them for death, while our national quotas on immigration precluded any addition of their kind. These two stories of past exclusion and truncated present welcome surely illustrate the familiar historical dictum that significant events tend to repeat themselves with an ironic difference—the first time as tragedy, the second as farce.

In 1925, Charles B. Davenport, one of America's foremost geneticists, wrote to his friend Madison Grant, the author of a best-selling book, *The Passing of the Great Race,* on the dilution of America's old (read northern European, not Indian) blood by recent immigration: "Our ancestors drove Baptists from Massachusetts Bay into Rhode Island, but we have no place to drive the Jews to." Davenport faced a dilemma. He sought a genetic argument for innate Jewish undesirability, but conventional stereotypes precluded the usual claim for inherent stupidity. So Davenport opted for weakness in moral character rather than intellect. He wrote in his 1911 book, *Heredity in Relation to Eugenics*—not, by the way, a political tract, but his generation's leading textbook in the developing science of genetics:

> In earning capacity both male and female Hebrew immigrants rank high and the literacy is above the mean of all immigrants. . . . On the other hand, they show the greatest proportion of offenses against chastity and in connection with prostitution . . . The hordes of Jews that are now coming to us from Russia and the extreme southeast of Europe, with their intense individualism and ideals of gain at the cost of any interest, represent the opposite extreme from the early English and the more recent Scandinavian immigration, with their ideals of community life in the open country, advancement by the sweat of the brow, and the uprearing of families in the fear of God and love of country.

The rediscovery and publication of Mendel's laws in 1900 initiated the modern study of genetics. Earlier theories of heredity had envisaged a "blending" or smooth mixture and dilution of traits by interbreeding with partners of different constitution, whereas Mendelism featured a "particulate" theory of inheritance, with traits coded by discrete and unchanging genes that need not be expressed in all offspring (especially if "recessive" to a "dominant" form of the gene carried on the other chromosome of a given pair), but that remain in the hereditary constitution, independent and undiluted, awaiting expression in some future generation.

In an understandable initial enthusiasm for this great discovery, early geneticists committed their most common and consistent error in trying to identify single genes as causes for nearly every human trait, from discrete bits of anatomy to complex facets of personality. The search for single genetic determinants seemed reasonable (and testable by analysis of pedigrees) for simple, discrete, and discontinuous characters and contrasts (like blue versus brown eyes). But the notion that complex behaviors and temperaments might also emerge from a similar root in simple heredity of single genes never made much sense, for two major reasons: (1) a continuity in expression that precludes any easy definition of traits supposedly under analysis (I may know blue eyes when I see them, but where does a sanguine personality end and melancholia take over?); and (2) a virtual certainty that environments can substantially mold such characters, whatever their underlying genetic influence (my eyes may become blue whatever I eat, but my inherently good brain may end up residing in a stupid adult if poor nutrition starved my early growth, and crushing poverty denied me an education).

Nonetheless, most early human geneticists searched for "unit characters"—supposed traits that could be interpreted as the product of a single Mendelian factor—with abandon, even in complex, continuous, environmentally labile, and virtually undefinable features of personality or accomplishment in life. (These early analyses proceeded primarily by the tracing of pedigrees. I can envisage accurate data, and reliable results, for a family chart of eye color, but how could anyone trace the alleged gene for "optimism," "feeble inhibition," or "wanderlust"—not to mention such largely situational phenomena as "pauperism" or "communality"? Was Great-uncle George a jovial backslapper or a reclusive cuss?)

Whatever the dubious validity of such overextended attempts to reduce complex human behaviors to effects of single genes, this strategy certainly served the aims and purposes of the early twentieth century's most influential social crusade with an allegedly scientific foundation: the eugenics movement, with its stated aim of "improving" America's hereditary stock by preventing procreation among the supposedly unfit (called "negative eugenics") and encouraging more breeding among those deemed superior in bloodline ("positive eugenics"). The abuses of this movement have been extensively documented in many excellent books covering such subjects as the hereditarian theory of mental testing, and the passage of legislation for involuntary sterilization and restriction of immigration from nations deemed inferior in hereditary stock.

Many early geneticists played an active role in the eugenics movement, but none more zealously than the aforementioned Charles Benedict Davenport (1866–1944), who received a Ph.D. in zoology at Harvard in 1892, taught at the University of Chicago, and then became head of the Carnegie Institution's Station for Experimental Evolution at Cold Spring Harbor, New York, where he also established and directed the Eugenics Record Office, beginning in

1910. This office, with mixed aims of supposedly scientific documentation and overt political advocacy, existed primarily to establish and compile detailed pedigrees in attempts to identify the hereditary basis of human traits. The hyperenthusiastic Davenport secured funding from several of America's leading (and in their own judgment, therefore eugenically blessed) families, particularly from Mrs. E. H. Harriman, the guardian angel and chief moneybags for the entire movement.

In his 1911 textbook, dedicated to Mrs. Harriman "in recognition of the generous assistance she has given to research in eugenics," Davenport stressed the dependence of effective eugenics upon the new Mendelian "knowledge" that complex behavioral traits may be caused by single genes. Writing of the five thousand immigrants who passed through Ellis Island every day, Davenport stated:

> Every one of these peasants, each item of that "riff-raff" of Europe, as it is sometimes carelessly called, will, if fecund, play a role for better or worse in the future history of this nation. Formerly, when we believed that factors blend, a characteristic in the germ plasm of a single individual among thousands seemed not worth considering: it would soon be lost in the melting pot. But now we know that unit characters do not blend; that after a score of generations the given characteristic may still appear, unaffected by repeated unions. . . . So the individual, as the bearer of a potentially immortal germ plasm with innumerable traits, becomes of the greatest interest.

—that is, of *our* "greatest interest" to exclude by vetting and restricting immigration, lest American heredity be overwhelmed with a deluge of permanent bad genes from the wretched refuse of foreign lands.

To illustrate Davenport's characteristic style of argument, and to exemplify his easy slippage between supposed scientific documentation and overt political advocacy, we may turn to his influential 1915 monograph entitled *The Feebly Inherited* (publication number 236 of his benefactors, the Carnegie Institute of Washington), especially to part 1 on "Nomadism, or The Wandering Impulse, With Special Reference to Heredity." The preface makes no bones about either sponsorship or intent. With three of America's wealthiest and most conservative families on board, one could hardly expect disinterested neutrality toward the full range of possible results. The Carnegies had endowed the general show, while Davenport paid homage to specific patrons: "The cost of training the field-workers was met by Mrs. E. H. Harriman, founder and principal patron of the Eugenics Record Office, and Mr. John D. Rockefeller, who paid also the salaries of many of the field-workers."

Davenport's preface also boldly admits his political position and purposes. He wishes to establish "feeble inhibition" as a category of temperament leading to inferior morality. Such a formulation will provide a one-two punch for identification of the eugenically unfit—bad intellect *and* bad morals. According to Davenport, the genetic basis of intelligence had already been documented in numerous studies of the feebleminded. But eugenics now needed to codify the second major reason for excluding immigrants and discouraging or denying reproductive rights to the native unfit—bad moral character (as in Davenport's fallback position, documented earlier in this essay, for restricting Jewish immigration when he could not invoke the usual charge of intellectual inferiority). Davenport writes:

> A word may be said as to the term "feebly inhibited" used in these studies. It was selected as a fit term to stand as co-ordinate with "feeble-minded" and as the result of a conviction that the phenomena with which it deals should properly be considered apart from those of feeble-mindedness.

To allay any doubt about his motivations, Davenport then makes his political point up front. Feeble inhibition, leading to immorality, may be more dangerous than feeblemindedness, leading to stupidity:

> I think it helps to consider separately the hereditary basis of the intellect and the emotions. It is in this conviction that these studies are submitted for thoughtful consideration. For, after all, the chief problem in administering society is that of disordered conduct, conduct is controlled by emotions, and the quality of the emotions is strongly tinged by the hereditary constitution.

Davenport then selects "nomadism" as his primary example of a putatively simple Mendelian trait—the product of a single gene—based on "feeble inhibition" and leading almost inevitably to immoral behavior. He encounters a problem of definition at the very outset of his work, as expressed in an opening sentence that must be ranked as one of the least profound in the entire history of science! "A tendency to wander in some degree is a normal characteristic of man, as indeed of most animals, in sharp contrast to most plants."

How then shall the "bad" form of wanderlust, defined as a compulsion to flee from responsibility, be distinguished from the meritorious sense of bravery and adventure—leading to "good" wanderlust—that motivated our early (and largely northern European) immigrants to colonize and subdue the frontier? Davenport had warmly praised the "good" form in his 1911 book as "the enterprising restlessness of the early settlers . . . the ambitious search for better conditions. The abandoned farms of New England point to the trait in our blood that entices us to move on to reap a possible advantage elsewhere."

In a feeble attempt to put false labels on segments of complex continua, Davenport identified the "bad" form as "nomadism," defined as an inability to inhibit the urge we all feel (from time to time) to flee from our duties, but that folks of normal and decent morality suppress. Nomads are society's tramps, bums, hoboes, and gypsies—"those who, while capable of steady and effective work, at more or less regular periods run away from the place where their duties lie and travel considerable distances."

Having defined his quarry (albeit in a fatally subjective way), Davenport then required two further arguments to make his favored link of a "bad" trait (rooted in feeble inhibition and leading to immoral behavior) to a single gene that eugenics might labor to breed down and out: he needed to prove the hereditary basis, and then to find the "gene," for nomadism.

His arguments for a genetic basis must be judged as astonishingly weak, even by the standards of his own generation (and despite the renown of his work, attributable, we must assume in retrospect, to its consonance with what most readers wanted to believe rather than to the quality of Davenport's logic or data). He simply argued, based on four dubious analogies, that features akin to nomadism emerge whenever situations veer toward "raw" nature (where genetics must rule), and away from environmental refinements of modern human society. Nomadism must be genetic because analogous features appear as "the wandering instinct in great apes," "among primitive peoples," in children (then regarded as akin to primitives under the false view that ontogeny recapitulates phylogeny), and in adolescents (where raw instinct temporarily overwhelms social inhibition in the *Sturm and Drang* of growing up). The argument about "primitive" people seems particularly weak, since a propensity for wandering might be regarded as well suited to a lifestyle based on hunting mobile game, rather than identified as a mark of inadequate genetic constitution (or any kind of genetic constitution at all). But Davenport, reversing the probable route of cause and effect, pushed through any difficulty to his desired conclusion:

> If we regard the Fuegians, Australians, Bushmen and Hottentots as the most primitive men, then we may say that primitive man is nomadic. . . . It is frequently assumed that they are nomadic because they hunt, but it is more probable that their nomadic instincts force them to hunting rather than agriculture for a livelihood.

Davenport then pursues his second claim—nomadism as the product of a single gene—by tracing pedigrees stored in his Eugenics Record Office. On the subjective criterion of impressions recorded by fieldworkers, or written descriptions of amateur informants (mostly people who had submitted their family trees in response to a general appeal for data), Davenport marked all nomads in his table with a scarlet *W* (for *Wanderlust,* the common German term for "urge to roam"). He then examined the distribution of *W*'s through families and generations to reach one of the most peculiar and improbable conclusions ever advanced in a famous study: nomadism, he argued, is caused by *a single gene,* a sex-linked recessive located on what would later be identified as the female chromosome.

Davenport reached this conclusion by arguing that nomadism occurred in families with the same distribution as hemophilia, colorblindness, and other truly sex-linked recessive traits. Such a status can be legitimately inferred from several definite patterns of heredity. For example, fathers with the trait do not pass it to their sons (since the relevant gene resides on the X-chromosome and males only pass a Y-chromosome to their sons). Mothers with the trait pass it to all their sons, but none of their daughters when the father lacks the trait. (Since the feature is recessive, an afflicted mother must carry the gene on both X-chromosomes. She passes a single X to her son, who must then express the trait, for he has no other X-chromosome. But a daughter will receive one afflicted X-chromosome from her mother and one normal X-chromosome from her father; she will therefore not express the trait because her father's normal copy of the gene is dominant.) Davenport knew these rules, so his study didn't fail on this score. Rather, his criteria for identifying "nomadism" as a discrete and scorable "thing" remained so subjective, and so biased by his genetic assumptions, that his pedigree data can only be judged as worthless.

Davenport's summary reached (and preached) a eugenic crescendo: "The wandering instinct," he stated, "is a fundamental human instinct, which is, however, typically inhibited in intelligent adults of civilized peoples." Unfortunately, however, people who express the bad gene *W* (the scarlet letter of wanderlust) cannot achieve this healthy inhibition, and become feckless nomads who run from responsibility by literal flight. The trait is genetic, racial, and undesirable. Immigrants marked by *W* should be excluded (and many immigrants must be shiftless wanderers rather than brave adventurers), while nomadic natives should be strongly encouraged, if not compelled, to desist from breeding. Davenport concludes:

> The new light brought by our studies is this: The nomadic impulse is, in all the cases, one and the same unit character. Nomads, of all kinds, have a special racial trait—are, in a proper sense, members of the nomadic race. This trait is the absence of the germinal determiner that makes for sedentariness, stability, domesticity.

Of course, no one would now defend Davenport's extreme view that single genes determine nearly every complex human behavior. Most colleagues eventually rejected Davenport's theory during his own career, especially since he lived into the 1940s, long past the early flush of Mendelian enthusiasm, and well past our modern recognition that complex traits usually record the operation of many genes, each with a small and cumulative effect (not to mention a strong, and often predominant influence from nongenetic environmental contexts of growth and expression). A single gene for anger, conviviality, contemplation, or wanderlust now seems as absurd as a claim that one assassin's bullet, and nothing else, caused World War I, or that Darwin discovered evolution all by himself, and we would still be creationists if he had never been born.

Nonetheless, in our modern age of renewed propensity for genetic explanations (a valid and genuine enthusiasm when properly pursued), Davenport's general style of error resurfaces on an almost daily basis, albeit in much more subtle form, but with all the vigor of his putative old gene—yes, he did propose one—for stubbornly persistent behavior.

No sensible critic of biological determinism denies that genes influence behavior; of course they do. Moreover, no honorable skeptic would argue that genetic explanations should be resisted because they entail negative political, social, or ethical connotations—a charge that must be rejected for two primary reasons. First, nature's facts stand neutral before our ethical usages. We have, to be sure, often made dubious, even tragic decisions based on false genetic claims. But in other contexts, valid arguments about the innate and hereditary basis of human attributes can be profoundly liberating. Consider only the burden lifted from loving parents who raise beautiful and promising children for twenty years, and then "lose" them to the growing ravages of schizophrenia—almost surely a genetically based disease of the mind, just as many congenital diseases of other bodily organs also appear in the third decade of life or even later. Generations of psychologists had subtly blamed parents for unintentionally inducing such a condition, then viewed as entirely "environmental" in origin. What could be more cruel than a false weight of blame added to such an ultimate tragedy? Second, we will never get very far, either in our moral deliberations or in our scientific inquiries, if we disregard genuine facts because we dislike their implications. In the most obvious case, I cannot think of a more unpleasant fact than the inevitable physical death of each human body, but no sane person would bet on extended stability for a society built on the premise that King Prospero will reign in his personal flesh forever.

However, if we often follow erroneous but deeply rooted habits of thinking to generate false conclusions about the role of heredity in human behavior, then these habits should be exposed and corrected—all the more vigorously if such arguments usually lead to recommendations for action that most people would also regard as ethically wrong (involuntary sterilization of the mentally retarded, for example). I believe that we face such a situation today, and that the genetic fallacies underlying our misusages bear a striking similarity in style and logic to Davenport's errors, however much we have gained in subtlety of argument and factual accuracy.

Throughout the history of genetics, political misuse has most frequently originated from claims for "biological determinism"—the argument that a given behavior or social situation can't be altered because people have been "built that way" by their genes. Once we attribute something we don't like to "genes," we tend either to make excuses, or to make less effort for change. In the most obvious, egregious, and persisting example, many people still argue that we should deny educational benefits and social services to groups (usually races or social classes) falsely judged as genetically inferior on average, because their poverty and misfortune lie in their own heredity and cannot be significantly ameliorated by social intervention. Thus, history shows a consistent linkage between genetic claims cast in this mold and conservative political arguments for maintenance of an unjust status quo of great benefit to people currently in power.

Of course, no serious student of either genetics or politics would now advance this argument in Davenport's style of "one gene, one complex behavior." That is, no one talks today about *the* gene for stupidity, promiscuity, or lack of ambition. But a series of three subtle—and extremely common—errors lead all too often to the same eugenical style of conclusion. Somehow we remain fascinated with the idea that complex social behaviors might be explained, at least in large part, by inherited "atoms" of behavioral propensity lying deeply within individuals. We seem so much more satisfied, so much more intrigued, by the claim that a definite gene, rather than a complex and inextricable mix of heredity and social circumstances, causes a particular phenomenon. We feel that we have come much nearer to a real or essential cause when we implicate a particle within an individual, rather than a social circumstance built of multiple components, as the reason behind a puzzling behavior. We will avidly read a front-page headline entitled "gay gene found," but newspapers will not even bother to report an equally well documented story on other components of homosexual preference with a primary social root and no correlated genetic difference.

The common source of these errors lies much deeper than any crude correlation to a political utility that most of us do not even recognize and would disavow if we did. The source lies, I believe, in a general view about causality that has either been beaten into us by a false philosophy about science and the natural world, or may even record an unfortunate foible in our brain's evolved mode of operation. We favor simple kinds of explanations that flow in one direction from small, independent, constituent atoms of being, to complex and messy interactions among large bodies or organizations. In other words, and to use the technical term, we prefer to be "reductionists" in our causal schemes—to explain the physical behavior of large objects as consequences of atoms in motion, or to explain the social behavior of large animals by biological atoms called genes.

But the world rarely matches our simplistic hopes, and the admittedly powerful methods of reductionism don't always apply. Wholes *can* be bigger than the sums of their parts, and interactions among objects cannot always be disaggregated into rules of action for each object considered separately. The rules and randomnesses of particular situations must often be inferred from direct and overt study of large objects and their interactions, not by reduction to constituent "atoms" and their fundamental properties. The three common errors of genetic explanation all share the same basic fallacy of reductionist assumptions.

1. We regard ourselves as sophisticated when we acknowledge that *both* genes and environment produce a given outcome, but we err in assuming that we can best express such a correct principle by assigning percentages and

stating, for example, that "behavior A is 40 percent genetic and 60 percent environmental." Such reductionist expressions pass beyond the status of simple error into the even more negative domain of entirely meaningless statements. Genetics and environment do interact to build a totality, but we need to understand why resulting wholes are unbreakable and irreducible to separate components. Water cannot be explained as two-thirds of the separate properties of hydrogen gas mixed with one-third of oxygen's independent traits—just as wanderlust cannot be analyzed as 30 percent of a gene for feeble inhibition mixed with 70 percent of social circumstances that abet an urge to hit the road.

2. We think that we have reached some form of subtle accuracy in saying that many genes, not just a Davenportian unity, set the hereditary basis of complex behaviors. But we then take this correct statement and reintroduce the central error of reductionism by asserting that if 10 genes influence behavior A, and if the causes of A may be regarded as 50 percent genetic (the first error), then each gene must contribute roughly 5 percent to the totality of behavior A. But complex interactions cannot be calculated as the sum of independent parts considered separately. I cannot be understood as one-eighth of each of my great-grandparents (though my genetic composition may be roughly so determined); I am a unique product of my own interactive circumstances of social setting, heredity composition, and all the slings and arrows of individual and outrageous natural fortune.

3. We suppose that we have introduced sufficient caution in qualifying statements about "genes for" traits by admitting their only partial, and often quite small, contribution to an interactive totality. Thus, we imagine that we may legitimately talk of a "gay gene" so long as we add the proviso that only 15 percent of sexual preference records this cause. But we need to understand why such statements have no meaning and therefore become (as for the first argument above) worse than merely false. Many genes interact with several other factors to influence sexual preference, but no unitary and separable "gay gene" exists. When we talk about a "gene for" 10 percent of behavior A, we simply commit the old Davenportian fallacy on the "little bit pregnant" analogy.

As a concrete example of how a good and important study can be saddled with all these errors in public reporting (and also by less than optimally careful statements of some participating researchers), *The New York Times* greeted 1996 with a headline on the front page of its issue for January 2: "Variant Gene Tied to a Love of New Thrills." The article discussed two studies published in the

January 1996 issue of *Nature Genetics.* Two independent groups of researchers, one working with 124 Ashkenazi and Sephardic Jews from Israel, the other with a largely male sample of 315 ethnically diverse Americans, both found a clearly significant, if weak, association between propensity for "novelty-seeking" behavior (as ascertained from standard survey questionnaires) and possession of a variant of a gene called the D4 dopamine receptor, located on the eleventh chromosome, and acting as one of at least five receptors known to influence the brain's response to dopamine.

This gene exists in several forms, defined by differing lengths recording the number (anywhere from two to ten) of repeated copies of a particular DNA subunit within the gene. Individuals with a high number of repeated copies (that is, with a longer gene) tended to manifest a greater tendency for "novelty-seeking" behavior—perhaps because the longer form of the gene somehow acts to enhance the brain's response to dopamine.

So far, so good—and very interesting. We can scarcely doubt that heredity influences broad and basic aspects of temperament—a bit of folk wisdom that surely falls into the category of "what every parent with more than one child knows." No one should feel at all offended or threatened by the obvious fact that we are not all born entirely blank, or entirely the same, in our mixture of the broad behavioral propensities defining what we call "temperament." Certain genes evidently influence particular aspects of brain chemistry; and brain chemistry surely affects our moods and behaviors. We know that basic and powerful neurotransmitters like dopamine strongly impact our moods and feelings (particularly, for dopamine, our sensations of pleasure). Differing forms of genes that affect the brain's response to dopamine may influence our behaviors—and a form that enhances the response may well incline a person toward "novelty-seeking" activities.

But the long form of the D4 receptor does not therefore become *the* (or even a) "novelty-seeking" gene, and these studies do not show that novelty seeking can be quantified and explained as a specified percent "genetic" in origin—although statements in this form dominated popular reports of these discoveries. Even the primary sources—the two original reports in *Nature Genetics* and the accompanying editorial feature entitled "Mapping Genes for Human Personality"—and the excellent *Times* story (representing the best of our serious press) managed, amidst their generally careful and accurate accounts, to propagate all three errors detailed above.

The *Times* reporter committed the first error of assigning separable percentages by writing "that about half of novelty-seeking behavior is attributable to genes, the other half to as yet ill defined environmental circumstances." Dr. R.P. Ebstein, principal author of one of the reports, then stated the second error of adding up effects without considering interactions when he argued that the long form of the D4 gene accounts for only about 10 percent of novelty-seeking behavior. If, by the first error, the totality of novelty seeking can be viewed as 50 percent genetic, and if D4 accounts for 10 percent of the totality, then we can infer that about four other genes must be involved (each contributing its 10 percent for the grand total of 50 percent genetic influence). Ebstein told the *Times* reporter: "If we assume that there are other genes out there that we haven't looked at yet, and that each gene exerts more or less the same influence as the D4 receptor, then we would expect maybe four or five genes are involved in the trait."

But the most significant errors, as always, fall into the third category of misproclaiming "genes for" specific behaviors—as in the title of the technical report from *Nature Genetics,* previously cited: "Mapping Genes for Human Personality." (If our professional journals so indulge and err, imagine what the popular press makes of "gay genes," "thrill genes," "stupidity genes," and so on.) First of all, the D4 gene by itself exerts only a weak potential influence upon novelty-seeking behavior. How can a gene accounting for only 10 percent of the variance in a trait be proclaimed as a "gene for" the trait? If I decide that 10 percent of my weight gain originated from the calories in tofu (because I love the stuff and eat it by the ton), this item, generally regarded as nutritionally benign, does not become a "fatness food."

More importantly, genes make enzymes, and enzymes control the rates of chemical processes. Genes do not make "novelty-seeking" or any other complex and overt behavior. Predisposition via a long chain of complex chemical reactions, mediated through a more complex series of life's circumstances, does not equal identification or even causation. At most, the long form of D4 induces a chemical reaction that can, among other possible effects, generate a mood leading some people to greater openness toward behaviors defined by some questionnaires as "novelty seeking."

In fact, a further study, published in 1997, illustrated this error in a dramatic way by linking the same long form of D4 to greater propensity for heroin addiction. The original *Times* article of 1996 had celebrated the "first known report of a link between a specific gene and a specific normal personality trait." But now the same gene—perhaps via the same route of enhanced dopamine response—also correlates with a severe pathology in other personalities. So what shall we call D4—a "novelty-seeking" gene in normal folk, or an "addiction" gene in troubled people? We need instead to reform both our terminology and our concepts. The long form of D4 induces a chemical response. This response may correlate with many different overt behaviors in people with widely varying histories and genetic constitutions.

The deepest error of this third category lies in the reductionist, and really rather silly, notion that we can even define discrete, separable, specific traits within the complex continua of human behaviors. We encounter enough difficulty in trying to identify characters with clear links to particular genes in the much clearer and simpler features of human anatomy. I may be able to specify genes "for" eye color, but not for leg length or fatness. How then shall I parse the continuous and necessarily subjective categories of labile personalities? Is "novelty seeking" really a "thing" at all? Can I even talk in a meaningful way about "genes for" such nebulous categories? Have I not fallen right back into the errors of Davenport's search for the internal scarlet letter *W* of wanderlust?

I finally realized what had been troubling me so much about the literature on "genes for" behavior when I read the *Times*'s account of C. R. Cloninger's theory of personality (Cloninger served as principal author of the *Nature Genetics* editorial commentary):

> Novelty seeking is one of four aspects that Dr. Cloninger and many other psychologists propose as the basic bricks of normal temperament, the other three being avoidance of harm, reward dependence and persistence. All four humors are thought to be attributable in good part to one's genetic makeup.

The last line crystallized my distress—"all four humors"—for I grasped, with the emotional jolt that occurs when all the previously unconnected pieces of an argument fall suddenly into place, why the canny reporter (or the scientist himself) had used this old word. Consider the theory in outline: four independent components of temperament, properly in balance in "normal" folks, with each individual displaying subtly different proportions, thus determining our individual temperaments and building our distinct personalities. But if our body secretes too much, or too little, of any particular humor, then a pathology may develop.

But why four, and why these four? Why not five, or six, or six hundred? Why any specific number? Why try to parse such continua into definite independent "things" at all? I do understand the mathematical theories and procedures that lead to such identifications (see my book *The Mismeasure of Man*), but

I regard the entire enterprise as a major philosophical error of our time (while I view the mathematical techniques, which I use extensively in my own research, as highly valuable when properly applied). Numerical clumps do not identify physical realities. A four-component model of temperament may act as a useful heuristic device, but I don't believe for a moment that four homunculi labeled *novelty seeking, avoidance of harm, reward dependence,* and *persistence* reside in my brain, either vying for dominance or cooperating in balance.

The logic of such a theory runs in uncanny parallel—hence the clever choice of "humor" as a descriptive term for the proposed modules of temperament—with the oldest and most venerable of gloriously wrong theories in the history of medicine. For more than a thousand years, from Galen to the dawn of modern medicine, prevailing wisdom regarded the human personality as a balance among four humors—blood, phlegm, choler, and melancholy. *Humor,* from the Latin word for "liquid" (a meaning still preserved in designating the fluids of the human eye as aqueous and vitreous humor), referred to the four liquids that supposedly formed the chyle, or digested food in the intestine just before it entered the body for nourishment. Since the chyle formed, on one hand, from a range of choices in the food we eat and, on the other hand, from constitutional differences in how various bodies digest this food, the totality recorded both innate and external factors—an exact equivalent to the modern claim that both genes and environment influence our behavior.

The four humors of the chyle correspond to the four possible categories of a double dichotomy—that is, to two axes of distinction based on warm–cold and wet–dry. The warm and wet humor forms blood; cold and wet generates phlegm; warm and dry makes choler; while cold and dry builds melancholy. I regard such a logically abstract scheme as a heuristic organizing device, much like Cloninger's quadripartite theory of personality. But we make a major error if we elevate such a scheme to a claim for real and distinct physical entities inside the body.

In the medical theory of humors, good health results from a proper balance among the four, while distinctive personalities emerge from different proportions within the normal range. But too much of any one humor may lead to oddness or pathology. As a fascinating linguistic remnant, we still use the names of all four humors as adjectives for types of personality—sanguine (dominance of the hot–wet blood humor) for cheerful people, phlegmatic for stolid folks dominated by the cold–wet humor of phlegm, choleric for angry individuals saddled with too much hot–dry choler, and melancholic for sad people overdosed with black bile, the cold–dry humor of melancholia. Does the modern quadripartite theory of personality really differ in any substantial way from this older view in basic concepts of number, balance, and the causes of both normal personality and pathology?

In conclusion, we might imagine two possible reasons for such uncanny similarity between a modern conception of four components to temperament, and the old medical theory of humors. Perhaps the similarity exists because the ancients had made a great and truthful discovery, while the modern version represents a major refinement of a central fact that our ancestors could only glimpse through a glass darkly. But alternatively—and ever so much more likely in my judgment—the stunning similarities exist because the human mind has remained constant throughout historical time, despite all our growth of learning and all the tumultuous changes in Western culture. We therefore remain sorely tempted by the same easy fallacies of reasoning.

I suspect that we once chose four humors, and now designate four end members of temperament, because something deep in the human psyche leads us to impose simple taxonomic schemes of distinct categories upon the world's truly complex continua. After all, our forebears didn't invoke the number four only for humors. We parsed many other phenomena into schemes with four end members—four compass points, four ages of man, and four Greek elements of air, earth, fire, and water. Could these similarities of human ordering be coincidental, or does the operation of the human brain favor such artificial divisions? Carl G. Jung, for reasons that I do not fully accept, strongly felt that division by four represented something deep and archetypal in human proclivities. He argued that we inherently view divisions by three as incomplete and leading onward (for one triad presupposes another for contrast), whereas divisions by four stand in optimal harmony and internal balance. He wrote: "Between the three and the four there exists the primary opposition of male and female, but whereas fourness is a symbol of wholeness, threeness is not."

I think that Jung correctly discerned an inherent mental attraction to divisions by four, but I suspect that the true basis for this propensity lies in our clear (and probably universal) preference for dichotomous divisions. Division by four may denote an ultimate and completed dichotomization—a dichotomy of dichotomies: two axes (each with two end members) at right angles to each other. We may experience four as an ultimate balance because such schemes fill our mental space with two favored dichotomies in perfect and opposite coordination.

In any case, if this second reason explains why we invented such eerily similar theories as four bodily humors and four end members of temperament, then such quadripartite divisions reflect biases of the mind's organization, not "real things" out there in the physical world. We can hardly talk about "genes for" the components of such artificial and prejudicial parsings of a much more complex reality. Interestingly, the greatest literary work ever written on the theory of humors, the early-seventeenth-century *Anatomy of Melancholy* by the English divine and scholar Robert Burton, properly recognized the four humors as just one manifestation of a larger propensity to divide by four. This great man who used the balm of literature to assuage his own lifelong depression, wrote of his condition: "Melancholy, cold and drie, thicke, blacke, and sowre . . . is a bridle to the other two hot humors, bloode and choler, preserving them in the blood, and nourishing the bones: These foure humors have some analogie with the foure elements, and to the foure ages in man."

I would therefore end—and where could an essayist possibly find a more appropriate culmination—with some wise words from Montaigne, the sixteenth-century founder of the essay as a literary genre. Perhaps we should abandon our falsely conceived and chimerical search for a propensity to wander, or to seek novelty (perhaps a spur to wandering), in a specific innate sequence of genetic coding. Perhaps, instead, we should pay more attention to the wondrous wanderings of our mind. For until we grasp the biases and propensities of our own thinking, we will never see through the humors of our vision into the workings of nature beyond. Montaigne wrote:

> It is a thorny undertaking, and more so than it seems, to follow a movement so wandering as that of our mind, to penetrate the opaque depths of its innermost folds, to pick out and immobilize the innumerable flutterings that agitate it.

19

Dolly's Fashion and Louis's Passion

NOTHING CAN BE MORE FLEETING OR CAPRICIOUS THAN fashion. What, then, can a scientist, committed to objective description and analysis, do with such a haphazardly moving target? In a classic approach, analogous to standard advice for preventing the spread of an evil agent ("kill it before it multiplies"), a scientist might say, "quantify it before it disappears."

Francis Galton, Charles Darwin's charmingly eccentric and brilliant cousin, and a founder of the science of statistics, surely took this prescription to heart. He once decided to measure the geographic patterning of female beauty. He therefore attached a piece of paper to a small wooden cross that he could carry, unobserved, in his pocket. He held the cross at one end in the palm of his hand and, with a needle secured between thumb and forefinger, made pinpricks on the three remaining projections (the two ends of the cross bar and the top). He would rank every young woman he passed on the street into one of three categories, as beautiful, average, or substandard (by his admittedly subjective preferences)—and he would then place a pinprick for each woman into the designated domain of this cross. After a hard day's work, he tabulated the relative percentages by counting pinpricks. He concluded, to the dismay of Scotland, that beauty followed a simple trend from north to south, with the highest proportion of uglies in Aberdeen, and the greatest frequency of lovelies in London.

Some fashions (body piercings, perhaps?) flower once and then disappear, hopefully forever. Others swing in and out of style, as if fastened to the end of a pendulum. Two foibles of human life strongly promote this oscillatory mode. First, our need to create order in a complex world, begets our worst mental habit: dichotomy (see chapter 3), or our tendency to reduce a truly intricate set of subtle shadings to a choice between two diametrically opposed alternatives (each with moral weight and therefore ripe for bombast and pontification, if not outright warfare): religion versus science, liberal versus conservative, plain versus fancy, "Roll Over Beethoven" versus the "Moonlight" Sonata. Second, many deep questions about our loves and livelihood, and the fates of nations, truly have no answers—so we cycle the presumed alternatives of our dichotomies, one after the other, always hoping that, this time, we will find the nonexistent key to an elusive solution.

Among oscillating fashions governed primarily by the swing of our social pendulum, no issue can claim greater prominence for an evolutionary biologist, or hold more relevance to a broad range of political questions, than genetic versus environmental sources of human abilities and behaviors. This issue has been falsely dichotomized for so many centuries that English even features a mellifluous linguistic contrast for the supposed alternatives: nature versus nurture.

As any thoughtful person understands, the framing of this question as an either-or dichotomy verges on the nonsensical. Both inheritance and upbringing matter in crucial ways. Moreover, an adult human being, built by interaction of these (and other) factors, cannot be disaggregated into separate components with attached percentages (see chapter 18 for detailed arguments on this vital issue). Nonetheless, a preference for either nature or nurture swings back and forth into fashion as political winds blow, and as scientific breakthroughs grant transient prominence to one or another feature in a spectrum of vital influences. For example, a combination of political and scientific factors favored an emphasis upon environment in the years just following World War II: an understanding that Hitlerian horrors had been rationalized by claptrap genetic theories about inferior races; the heyday of behaviorism in psychology.

Today genetic explanations have returned to great vogue, fostered by a similar mixture of social and scientific influences: a rightward shift of the political pendulum (and the cynical availability of "you can't change them, they're made that way" as a bogus argument for reducing government expenditures on social programs); and an overextension to all behavioral variation of genuinely exciting results in identifying the genetic basis of specific diseases, both physical and mental.

Unfortunately, in the heat of immediate enthusiasm, we often mistake transient fashion for permanent enlightenment. Thus, many people assume that the current popularity of genetic determinism represents a permanent truth, finally wrested from the clutches of benighted environmentalists of previous generations. But the lessons of history suggest that the worm will soon turn again. Since both nature and nurture can teach us so much—and since the fullness of our behavior and mentality represents such a complex and unbreakable combination of these and other factors—a current emphasis on nature will no doubt yield to a future fascination with nurture as we move toward better understanding by lurching upward from one side to another in our quest to fulfill the Socratic injunction: know thyself.

In my Galtonian desire to measure the extent of current fascination with genetic explanations (before the pendulum swings once again and my opportunity evaporates), I hasten to invoke two highly newsworthy items of recent times. The subjects may seem quite unrelated—Dolly the cloned sheep, and Frank Sulloway's book on the effects of birth order upon human behavior*—but both stories share a curious common feature offering striking insight into the current extent of genetic preferences. In short, both stories have been reported almost entirely in genetic terms, but both cry out (at least to me) for

*This essay, obviously, represents my reaction to the worldwide storm of news and ethical introspection launched by the public report of Dolly, the first mammal cloned from an adult cell, in early 1997. In collecting several years of essays together to make each of these books, I usually deep-six the rare articles keyed to immediate news items of "current events"—for the obvious reason of their transiency under the newsman's adage that "yesterday's paper wraps today's garbage." But in rereading this essay, I decided that it merited reprinting on two counts: first, I don't think that its relevance has at all faded (while Dolly herself also persists firmly in public memory); second, I fancy that I found something general and original to say by linking Dolly to Sulloway's book, and by relating both disparate events to a common theme that had puzzled me enormously by being so blessedly obvious, yet so totally unreported in both the serious and popular press. As King Lear discovered to his sorrow, the absence of an expected statement can often be far more meaningful than an anticipated and active pronouncement. Since these essays experience a three-month "lead time" between composition and original publication, I must always treat current events in a more general context potentially meriting republication down the line—for ordinary fast-breaking news can only become rock-hard stale in those interminable ninety days.

a radically different reading as proof of strong environmental influences. Yet no one seems to be drawing (or even mentioning) this glaringly obvious inference. I cannot imagine that anything beyond current fashion for genetic arguments can explain this puzzling silence. I am convinced that exactly the same information, if presented twenty years ago in a climate favoring explanations based on nurture, would have elicited a strikingly different interpretation. Our world, beset by ignorance and human nastiness, contains quite enough background darkness. Should we not let both beacons shine all the time?

CREATING SHEEP

Dolly must be the most famous sheep since John the Baptist designated Jesus in metaphor as "lamb of God, which taketh away the sin of the world" (John 1:29). She has certainly edged past the pope, the president, Madonna, and Michael Jordan as the best-known mammal of the moment. And all this brouhaha for a carbon copy, a Xerox! I don't mean to drip cold water on this little lamb, cloned from a mammary cell of her adult mother, but I remain unsure that she's worth all the fuss and fear generated by her unconventional birth.

When one reads the technical article describing Dolly's manufacture (I. Wilmut, A. E. Schnieke, J. McWhir, A. J. Kind, and K. H. S. Campbell, "Viable offspring derived from fetal and adult mammalian cells," *Nature,* February 27, 1997, pages 810–13), rather than the fumings and hyperbole of so much public commentary, one can't help feeling a bit underwhelmed, and left wondering whether Dolly's story tells less than meets the eye.

I don't mean to discount or underplay the ethical issues raised by Dolly's birth (and I shall return to this subject in a moment), but we are not about to face an army of Hitlers or even a Kentucky Derby run entirely by genetically identical contestants (a true test for skills of jockeys and trainers!). First, Dolly breaks no theoretical ground in biology, for we have known how to clone in principle for at least two decades, but had developed no techniques for reviving the full genetic potential of differentiated adult cells. (Still, I admit that a technological solution can pack as much practical and ethical punch as a theoretical breakthrough. I suppose one could argue that the first atomic bomb only realized a known possibility.)

Second, my colleagues have been able to clone animals from embryonic cell lines for several years, so Dolly does not rank as the first mammalian clone, but rather as the first clone from an adult cell. Ian Wilmut and his coworkers also cloned sheep from cells of a nine-day embryo and a twenty-six-day fetus—with much greater success. They achieved fifteen pregnancies (though not all proceeded to term) in thirty-two "recipients" (that is, surrogate mothers for transplanted cells) for the embryonic cell line, and five pregnancies in sixteen recipients for the fetal cell line, but only Dolly (one pregnancy in thirteen tries) for the adult cell line. This experiment cries out for confirming repetition. (Still, I allow that current difficulties will surely be overcome, and cloning from adult cells, if doable at all, will no doubt be achieved more routinely as techniques and familiarity improve.)*

Third, and more seriously, I remain unconvinced that we should regard Dolly's starting cell as adult in the usual sense of the term. Dolly grew from a cell taken from the "mammary gland of a 6-year-old ewe in the last trimester of pregnancy" (to quote the technical article of Wilmut et al.). Since the breasts of pregnant mammals enlarge substantially in late stages of pregnancy, some mammary cells, though technically adult, may remain unusually labile or even "embryolike," and thus able to proliferate rapidly to produce new breast tissue at an appropriate stage of pregnancy. Consequently, we may only be able to clone from unusual adult cells with effectively embryonic potential, and not from any stray cheek cell, hair follicle, or drop of blood that happens to fall into the clutches of a mad Xeroxer. Wilmut and colleagues admit this possibility in a sentence written with all the obtuseness of conventional scientific prose, and therefore almost universally missed by journalists: "We cannot exclude the possibility that there is a small proportion of relatively undifferentiated stem cells able to support regeneration of the mammary gland during pregnancy."

But if I remain relatively unimpressed by achievements thus far, I do not discount the monumental ethical issues raised by the possibility of cloning from adult cells. Yes, we have cloned fruit trees for decades by the ordinary process of grafting—and without raising any moral alarms. Yes, we may not face the evolutionary dangers of genetic uniformity in crop plants and livestock, for I trust that plant and animal breeders will not be stupid enough to eliminate all but one genotype from a species, and will always maintain (as plant breeders now do) an active pool of genetic diversity in reserve. (But then, I suppose we should never underestimate the potential extent of human stupidity—and

*Science moves fast, especially when spurred by immense public interest and pecuniary possibilities. These difficulties have been much mitigated, if not entirely overcome, in the three years between my original writing and this republication. Cloning from adult cells has not become, by any means, routine, but undoubted clones have been produced from adult cells in several mammalian species. Moreover, initial doubts about Dolly herself (mentioned in this essay) have largely been allayed, and her status as a clone from an adult cell now seems secure.

localized reserves could be destroyed by a catastrophe, while genetic diversity spread throughout a species guarantees maximal evolutionary robustness.)

Nonetheless, while I regard many widely expressed fears as exaggerated, I do worry deeply about potential abuses of human cloning, and I do urge an open and thorough debate on these issues. Each of us can devise a personal worst-case scenario. Somehow I do not focus upon the specter of a future Hitler making an army of 10 million identical robotic killers—for if our society ever reaches a state where someone in power could actually realize such an outcome, we are probably already lost. My thoughts run to localized moral quagmires that we might actually have to face in the next few years—for example, the biotech equivalent of ambulance-chasing slimeballs among lawyers: a hustling firm that scans the obits for reports of children who died young, and then goes to grieving parents with the following offer: "So sorry for your loss; but did you save a hair sample? We can make you another for a mere fifty thou."

However, and still on the subject of ethical conundrums, but now moving to my main point about current underplaying of environmental sources for human behaviors, I do think that the most potent scenarios of fear, and the most fretful ethical discussions of radio talk shows, have focused on a nonexistent problem that all human societies solved millennia ago. We ask: is a clone an individual? Would a clone have a soul? Would a clone made from my cell negate my unique personhood?

May I suggest that these endless questions—all variations on the theme that clones threaten our traditional concept of individuality—have already been answered empirically, even though public discussion of Dolly seems blithely oblivious to this evident fact.* We have known human clones from the dawn of our consciousness. We call them identical twins—and they constitute far better clones than Dolly and her mother. Dolly shares only nuclear DNA with her genetic mother—for only the nucleus of her mother's mammary cell was

*My deep puzzlement over public surprise at this obvious point, and at the failure of media to grasp and highlight the argument immediately, has only grown since I wrote this essay. (I believe that I was the first to stress or even to mention—in commentary to journalists before I published this article—the clonal nature of identical twins as an ancient and conclusive disproof of the major ethical fears that Dolly had so copiously inspired. No argument of such a basic and noncryptic nature should ever be first presented by a magazine essayist with a lead time of several months, rather than the next day by a journalist, or the next minute in cyberspace.) I can only conclude that public misunderstanding of environmental impact upon human personalities, emotions, and distinctivenesses runs much deeper than even I had realized, and that barriers to recognition of this self-evident truth stand even higher than I had suspected in the light of current fashions for genetic explanations.

inserted into an embryonic stem cell (whose own nucleus had been removed) of a surrogate female. Dolly then grew in the womb of this surrogate.

Identical twins share at least four additional (and important) attributes that differ between Dolly and her mother. First, identical twins also carry the same mitochondrial genes. (Mitochondria, the "energy factories" of cells, contain a small number of genes. We obtain our mitochondria from the cytoplasm of the egg cell that made us, not from the nucleus formed by the union of sperm and egg. Dolly received her nucleus from her mother but her egg cytoplasm, and hence her mitochondria, from her surrogate.) Second, identical twins share the same set of maternal gene products in the egg. Genes don't grow embryos all by themselves. Egg cells contain protein products of maternal genes that play a major role in directing the early development of the embryo. Dolly's embryology proceeded with her mother's nuclear genes but her surrogate's gene products in the cytoplasm of her founding cell.

Third—and now we come to explicitly environmental factors—identical twins share the same womb. Dolly and her mother gestated in different places. Fourth, identical twins share the same time and culture (even if they fall into the rare category, so cherished by researchers, of siblings separated at birth and raised, unbeknownst to each other, in distant families of different social classes). The clone of an adult cell matures in a different world. Does anyone seriously believe that a clone of Beethoven, grown today, would sit down one fine day to write a tenth symphony in the style of his early-nineteenth-century forebear?

So identical twins are truly eerie clones—much more alike on all counts than Dolly and her mother. We do know that identical twins share massive similarities, not only of appearance, but also in broad propensities and detailed quirks of personality. Nonetheless, have we ever doubted the personhood of each member in a pair of identical twins? Of course not. We know that identical twins are distinct individuals, albeit with peculiar and extensive similarities. We give them different names. They encounter divergent experiences and fates. Their lives wander along disparate paths of the world's complex vagaries. They grow up as distinctive and undoubted individuals, yet they stand forth as far better clones than Dolly and her mother.

Why have we overlooked this central principle in our fears about Dolly? Identical twins provide sturdy proof that inevitable differences of nurture guarantee the individuality and personhood of each human clone. And since any future human Dolly must depart far more from her progenitor (in both the nature of mitochondria and maternal gene products, and the nurture of different wombs and surrounding cultures) than any identical twin differs from her sibling clone, why ask if Dolly has a soul or an independent life when we have never doubted the personhood or individuality of far more similar identical twins?

Literature has always recognized this principle. The Nazi loyalists who cloned Hitler in *The Boys from Brazil* understood that they had to maximize similarities of nurture as well. So they fostered their little Hitler babies in families maximally like Adolf's own dysfunctional clan—and not one of them grew up anything like history's quintessential monster. Life has always verified this principle as well. Eng and Chang, the original Siamese twins and the (literally) closest clones of all, developed distinct and divergent personalities. One became a morose alcoholic, the other remained a benign and cheerful man. We may not attribute much individuality to sheep in general—they do, after all, establish our icon of blind following and identical form as they jump over fences in the mental schemes of insomniacs—but Dolly will grow up to be as unique and as ornery as any sheep can be.

KILLING KINGS

My friend Frank Sulloway recently published a book that he had fretted over, nurtured, massaged, and lovingly shepherded toward publication for more than two decades. Frank and I have been discussing his thesis ever since he began his studies. I thought (and suggested) that he should have published his results twenty years ago. I still hold this opinion—for while I greatly admire his book, and do recognize that such a long gestation allowed Frank to strengthen his case by gathering and refining his data, I also believe that he became too committed to his central thesis, and tried to extend his explanatory umbrella over too wide a range, with arguments that sometimes smack of special pleading and tortured logic.

Born to Rebel documents a crucial effect of birth order in shaping human personalities and styles of thinking. Firstborns, as sole recipients of parental attention until the arrival of later children, and as more powerful than their subsequent siblings (by virtue of age and size), generally cast their lot with parental authority and with the advantages of incumbent strength. They tend to grow up competent and confident, but also conservative and unlikely to favor quirkiness or innovation. Why threaten an existing structure that has always offered you clear advantages over siblings? Later children, however, are (as Sulloway's title proclaims) born to rebel. They must compete against odds for parental attention long focused primarily elsewhere. They must scrap and struggle, and learn to make do for themselves. Laterborns therefore tend to be flexible, innovative, and open to change. The business and political leaders of stable nations are usually firstborns, but the revolutionaries who have discombobulated our cultures and restructured our scientific knowledge tend to be laterborns.

Sulloway defends his thesis with statistical data on the relationship of birth order and professional achievements in modern societies, and by interpreting historical patterns as strongly influenced by characteristic differences in behaviors between firstborns and laterborns. I found some of his historical arguments fascinating and persuasive when applied to large samples (but often uncomfortably overinterpreted in trying to explain the intricate details of individual lives, for example the effect of birth order on the differential success of Henry VIII's various wives in overcoming his capricious cruelties).

In a fascinating case, Sulloway chronicles a consistent shift in relative percentages of firstborns among successive groups in power during the French Revolution. The moderates initially in charge tended to be firstborns. As the revolution became more radical, but still idealistic and open to innovation and free discussion, laterborns strongly predominated. But when control then passed to the uncompromising hard-liners who promulgated the Reign of Terror, firstborns again ruled the roost. In a brilliant stroke, Sulloway tabulates the birth orders for several hundred delegates who decided the fate of Louis XVI in the National Convention. Among hard-liners who voted for the guillotine, 73 percent were firstborns; but 62 percent of laterborns opted for the compromise of conviction with pardon. Since Louis lost his head by a margin of one vote, an ever-so-slightly different mix of birth orders among delegates might have altered the course of history.

Since Frank is a good friend (though I don't accept all details of his thesis), and since I have been at least a minor midwife to this project for two decades, I took an unusually strong interest in the delayed birth of *Born to Rebel.* I read the text and all the prominent reviews that appeared in many newspapers and journals. And I have been puzzled—*stunned* would not be too strong a word—by the total absence from all commentary of the simplest and most evident inference from Frank's data, the one glaringly obvious point that everyone should have stressed, given the long history of issues raised by such information.

Sulloway focuses nearly all his interpretation on an extended analogy (broadly valid in my judgment, but overextended as an exclusive device) between birth order in families and ecological status in a world of Darwinian competition. Children vie for limited parental resources, just as individuals struggle for existence (and ultimately for reproductive success) in nature. Birth order places children in different "niches," requiring disparate modes of competition for maximal success. While firstborns shore up incumbent advantages, laterborns must grope and grub by all clever means at their disposal—leading to the divergent personalities of stalwart and rebel. Alan Wolfe, in my favorite negative review in *The New Republic,* writes (December 23, 1996; Jared Diamond stresses the same themes in my favorite positive review in *The New York Review of Books,* November 14, 1996): "Since firstborns already occupy their own niches, laterborns, if they are to be noticed, have to find unoccupied niches. If they do so successfully, they will be rewarded with parental investment."

As I said, I am willing to follow this line of argument up to a point. But I must also note that restriction of commentary to this Darwinian metaphor has diverted attention from the foremost conclusion revealed by a large effect of birth order upon human behavior. The Darwinian metaphor smacks of biology; we also (albeit erroneously) tend to regard biological explanations as intrinsically genetic. I suppose that this common but fallacious chain of argument leads us to stress whatever we think that Sulloway's thesis might be teaching us about "nature" (our preference, in any case, during this age of transient fashion for genetic causes) under our erroneous tendency to treat the explanation of human behavior as a debate between nature and nurture.

But consider the meaning of birth-order effects for environmental influences, however unfashionable at the moment. Siblings differ genetically of course, but no aspect of this genetic variation correlates in any systematic way with birth order. Firstborns and laterborns receive the same genetic shake within a family. *Systematic* differences in behavior between firstborns and laterborns therefore cannot be ascribed to genetics. (Other biological effects may correlate with birth order—if, for example, the environment of the womb changes systematically with numbers of pregnancies—but such putative influences bear no relationship whatever to genetic differences among siblings.) Sulloway's substantial birth-order effect therefore provides our best and ultimate documentation of nurture's power. If birth order looms so large in setting the paths of history and the allocation of people to professions, then nurture cannot be denied a powerfully formative role in our intellectual and behavioral variation. To be sure, we often fail to see what stares us in the face; but how can the winds of fashion blow away such an obvious point, one so relevant to our deepest and most persistent questions about ourselves?

In this case, I am especially struck by the irony of fashion's veil. As noted above, I urged Sulloway to publish his data twenty years ago—when (in my judgment) he could have presented an even better case because he had already documented the strong and general influence of birth order upon personality, but had not yet ventured upon the slippery path of trying to explain too many details with forced arguments that sometimes lapse into self-parody. If Sulloway had published in the mid-1970s, when nurture rode the pendulum of fashion in a politically more liberal age (probably dominated by laterborns!), I am confident that this obvious argument about effects of birth order as proof of nurture's power would have won primary attention, rather than consignment to a limbo of invisibility.

Hardly anything in intellectual life can be more salutatory than the separation of fashion from fact. Always suspect fashion (especially when the moment's custom matches your personal predilection); always cherish fact (while remembering that an apparent jewel of pure and objective information may only record the biased vision of transient fashion). I have discussed two subjects that couldn't be "hotter," but cannot be adequately understood because a veil of genetic fashion now conceals the richness of full explanation by relegating a preeminent environmental theme to invisibility. Thus, we worry whether the first cloned sheep represents a genuine individual at all, while we forget that we have never doubted the distinct personhood guaranteed by differences in nurture to clones far more similar by nature than Dolly and her mother—identical twins. And we try to explain the strong effects of birth order solely by invoking a Darwinian analogy between family status and ecological niche, while forgetting that these systematic effects cannot arise from genetic differences, and can therefore only demonstrate the predictable power of nurture. Sorry, Louis. You lost your head to the power of family environments upon head children. And hello, Dolly. May we forever regulate your mode of manufacture, at least for humans. But may genetic custom never stale the infinite variety guaranteed by a lifetime of nurture in the intricate complexity of our natural world—this vale of tears, joy, and endless wonder.

20

Above All, Do No Harm

LONG, STAGNANT, AND COSTLY WARS TEND TO BEGIN in idealistic fervor and end in cynical misery. Our own Civil War inflicted a horrendous toll of death and seared our national consciousness with a brand that has only become deeper with time. In 1862, the Union Army rejoiced in singing the year's most popular ditty:

> *Yes we'll rally round the flag, boys, we'll rally once again*
> *Shouting the battle cry of Freedom,*
> *We will rally from the hillside, we'll gather from the plain,*
> *Shouting the battle cry of Freedom . . .*
> *So we're springing to the call from the East and from the West*
> *And we'll hurl the rebel crew from the land we love the best.*

By 1864, Walter Kittredge's "Tenting on the Old Camp Ground" had become the favorite song of both sides. The chorus, with its haunting (if naive) melody, summarizes the common trajectory:

Many are the hearts that are weary tonight,
Wishing for the war to cease;
Many are the hearts looking for the right
To see the dawn of peace.

But nothing can quite match the horrors of World War I, the conflict that the French still call *la grande guerre* (the Great War) and that we labeled "the war to end all wars." America entered late and suffered relatively few casualties as a consequence—so we rarely appreciate the extent of carnage among soldiers or the near certainty of death or serious maiming along lines of stagnant trenches, where men fought back and forth month after month to take, and then lose again, a few shifting feet of territory. I feel chills up and down my spine whenever I look at the "honor roll" posted on the village green or main square of any small town in England or France. Above all else, I note the much longer lists for 1914–18 (often marking the near extermination of a generation of males) than for 1941–45. Rupert Brooke could write his famous poems of resignation and patriotism because he died in 1915, during the initial blush of enthusiasm:

If I should die, think only this of me:
That there's some corner of a foreign field
That is for ever England. There shall be
In that rich earth a richer dust concealed.

An actual gas attack in World War I.

His fellow poet Siegfried Sassoon, who survived and became a pacifist (a condition first attributed to shell shock and leading to his temporary confinement in a sanatorium), caught the drift of later realism:

And when the war is done and youth stone dead
I'll toddle safely home and die—in bed.

Sassoon met Wilfred Owen, the third member of this famous trio of British war poets, in the sanatorium. But Owen went back to the front, where he fell exactly one week before Armistice Day. Sassoon published his friend's single, slim volume of posthumous poetry, containing the most famous and bitter lines of all:

What passing-bells for these who died as cattle?
Only the monstrous anger of the guns.
Only the stuttering rifles' rapid rattle
Can patter out their hasty orisons.

Among the horrors of World War I, we remember not only the carnage caused by conventional tactics of trench warfare with bombs and bullets but also the first effective and large-scale use of newfangled chemical and biological weapons—beginning with a German chlorine gas attack along four miles of the French line at Ypres on April 22, 1915, and ending with 100,000 tons of various chemical agents used by both sides. The Geneva Protocol, signed in 1925 by most major nations (but not by the United States until much later), banned both chemical and biological weapons—a prohibition followed by all sides in World War II, even amid some of the grimmest deeds in all human history, including, and let us never forget, the most evil acts ever committed with poisonous gases in executing the "final solution" of the Holocaust in Nazi concentration camps. (A few violations have occurred in local wars: by the Italian army in Ethiopia in 1935–36, for example, and in recent fighting in Iran and Iraq.) The Geneva Protocol prohibited "the use in war of asphyxiating, poisonous or other gases, and of all analogous liquids, materials or devices."

A recent contribution to *Nature* (June 25, 1998), the leading British professional journal of science, recalled this episode of twentieth-century history in a remarkable letter entitled "Deadly relic of the Great War." The opening paragraph reads:

> The curator of a police museum in Trondheim, Norway, recently discovered in his archive collection a glass bottle containing two irregularly shaped sugar lumps. A small hole had been bored into each of these lumps and a glass capillary tube, sealed at its tip, was embedded into one of the lumps. A note attached to the exhibit translated as follows: "A piece of sugar containing anthrax bacilli, found in the luggage of Baron Otto Karl von Rosen, when he was apprehended in Karasjok in January 1917, suspected of espionage and sabotage."

Modern science to the rescue, even in pursuit of a mad scheme that came to naught in a marginal and forgotten outpost of a great war—the very definition of historical trivia, however intriguing, in the midst of great pith and moment. The authors of the letter removed the capillary tube and dumped the contents ("a brown fluid") onto a petri dish. Two columns of conventional scientific prose then detailed the procedures followed, with all the usual rigor of long chemical names and precise amounts: "After incubation, 200μ1 of these cultures were spread on 7% of horse-blood agar and L-agar medium (identical to L-broth but solidified by the addition of 2% Difco Bacto agar)." The clear results may be stated more succinctly, as the authors both grew some anthrax bacilli in their cultures and then confirmed the presence of DNA from the same organism by PCR (polymerase chain reaction for amplifying small amounts of DNA to levels that can be analyzed). They write: "We therefore confirmed the presence of *B. anthracis* [scientific name of the anthrax bacillus] in the specimen by both culture and PCR. It proved possible to revive a few surviving organisms from the brink of extinction after they had been stored, without any special precautions, for 80 years."

But what was the good baron, an aristocrat of German, Swedish, and Finnish extraction, doing in this forsaken area of northeastern Norway in the middle of winter? Clearly up to no good, but to what form of no good? The authors continue:

> When the Sheriff of Kautokeino, who was present at the group's arrest, derisively suggested that he should prepare soup from the contents of the tin cans labeled "Svea kott" (Swedish meat), the baron felt obliged to admit that each can actually contained between 2 and 4 kilograms of dynamite.

The baron's luggage also yielded some bottles of curare, various microbial cultures, and nineteen sugar cubes, each containing anthrax. The two cubes in Trondheim are, apparently, the only survivors of this old incident. The baron claimed that he was only an honorable activist for Finnish independence, out to destroy supply lines to Russian-controlled areas. (Finland had been under loose control of the Russian czar and did win independence after the Bolshevik revolution.) Most historians suspect that he had traveled to Norway at the behest, and in the employ, of the Germans, who had authorized a program for infecting horses and reindeer with anthrax to disrupt the transport of British arms (on sleds pulled by these animals) through northern Norway.

The baron, expelled after a few weeks in custody, never carried out his harebrained scheme. The authors of the *Nature* letter, Caroline Redmond, Martin J. Pearce, Richard J. Manchee, and Bjorn P. Berdal, have inferred his intent:

> The grinding of the sugar and its glass insert between the molar teeth of horses would probably result in a lethal infection as the anthrax spores entered the body, eventually facilitated through the small lesions produced in the wall of the alimentary tract by the broken glass. It is not known whether reindeer eat sugar lumps but presumably the baron never had the chance to carry out this piece of research.

As anthrax cannot be transmitted directly from animal to animal, the scheme probably would not have worked without a large supply of sugar cubes and very sweet teeth in the intended victims. But the authors do cite a potential danger to other participants: "However, if the meat from a dying animal had been consumed without adequate cooking, it is likely that human fatalities from gastrointestinal anthrax would have followed." The authors end their letter with a frank admission:

> This small but relatively important episode in the history of biological warfare is one of the few instances where there is confirmation of the intent to use a lethal microorganism as a weapon, albeit 80 years after the event. It did not, however, make any significant difference to the course of the Great War.

We may treat this botched experiment in biological warfare as light relief in a dark time, but the greatest evils often begin as farcical and apparently harmless escapades, while an old motto cites eternal vigilance as the price of liberty. If Hitler had been quietly terminated after his ragtag band failed to seize local power in their Beer Hall Putsch of 1923 in Munich—even the name of this incident marks the derision then heaped on the protagonists—the history of our century might have unfolded in a much different, and almost surely happier, manner. Instead, Hitler spent a mere nine months in jail, where he wrote *Mein Kampf* and worked out his grisly plans.

We humans may be the smartest objects that ever came down the pike of life's history on earth, but we remain outstandingly inept in certain issues, particularly when our emotional arrogance joins forces with our intellectual ignorance. Our inability to forecast the future lies foremost among these ineptitudes—not, in this case, as a limitation of our brains, but more as a principled consequence of the world's genuine complexity and indeterminism (see chapter 10 for a general discussion of our inability to predict coming events and patterns). We could go with this flow, but our arrogance intercedes, leading us to promote our ignorant intuitions into surefire forecasts about things to come.

I know only one antidote to the major danger arising from this incendiary mixture of arrogance and ignorance. Given our inability to predict the future, particularly our frequent failures to forecast the later and dire consequences of phenomena that seem impotent, or even risible, at their first faltering steps (a few reindeer with anthrax today, an entire human population with plague tomorrow), moral restraint may represent our only sure salvation. The wisdom of the Geneva Protocol lies in understanding that some relatively ineffective novelties of 1925 might become the principal horrors of a not-so-distant future. If such novelties can be nipped in the bud of their early ineffectiveness, we may be spared. We need only remember the legend of Pandora to recognize that some boxes, once opened, cannot be closed again.

The good sense in this vital form of moral restraint has been most seriously and effectively challenged by scientists who stand at the cutting edge of a developing technology and therefore imagine that they can control, or at least accurately forecast, any future developments. I dwell in the camp of scientists, but I want to illustrate the value of moral restraint as a counterweight to dangerous pathways forged either by complacency or active pursuit and fueled by false confidence about forecasting the future.

I told a story about aristocratic bumbling with ineffective biological weapons in World War I—but we might be in quite a fix today if we had assumed that this technology could never transcend such early ineptitude, and if we had not worked hard for international restriction. But a much deeper lesson may be drawn from the second innovative, and much more effective, technology later banned by the Geneva Protocol: chemical weaponry in World War I. The primary figure for this lesson became one of the founders of my own field of modern evolutionary biology—J.B.S. Haldane (1892–1964), called "the cleverest man I ever knew" by Sir Peter Medawar, who was certainly the cleverest man I have ever known.

Haldane mixed so many apparently contradictory traits into his persona that one word stands out in every description of him I have ever read: *enigmatic.* He could be shy and kind or blustering and arrogant, elitist (and viciously dismissive of underlings who performed a task poorly) or egalitarian. (Haldane became a prominent member of the British Communist Party and wrote volumes of popular essays on scientific subjects for their *Daily Worker.* Friends, attributing his political views to a deep personal need for iconoclastic and contrarian behavior, said that he would surely have become a monarchist if he had lived in the Soviet Union.) Haldane held no formal degree in science but excelled in several fields, largely as a consequence of superior mathematical ability. He remains most famous, along with R. A. Fisher and Sewall Wright, as one of the three founders of the modern theory of population genetics,

J.B.S. Haldane in his World War I military outfit.

especially for integrating the previously warring concepts of Mendelian rules for heredity with Darwinian natural selection.

But a different contradiction motivates Haldane's appearance as the focus of this essay. Haldane, a man of peace and compassion, adored war—or at least his role on the front lines in World War I, where he was twice wounded (both times seriously) and mighty lucky to come home in one piece. Some people regarded him as utterly fearless and courageous beyond any possible call of duty; others, a bit more cynically perhaps (but also, I suspect, more realistically), viewed him as a latter-day Parsifal—a perfect fool who survived in situations of momentary danger (usually created as a result of his own bravado and appalling recklessness) by a combination of superior intelligence joined with more dumb luck than any man has a right to expect. In any case, J. B. S. Haldane had a good war—every last moment of it.

He particularly enjoyed a spell of trench warfare against Turkish troops near the Tigris River, where, away from the main European front and unencumbered by foolish orders from senior officers without local experience, men could fight *mano a mano* (or at least gun against gun). Haldane wrote: "Here men were pitted against individual enemies with similar weapons, trench mortars or rifles with telescopic sights, each with a small team helping him. This was war as the great poets have sung it. I am lucky to have experienced it." Haldane then offered a more general toast to such a manly occupation: "I enjoyed the comradeship of war. Men like war because it is the only socialized activity in which they have ever taken part. The soldier is working with comrades for a great cause (or so at least he believes). In peacetime he is working for his own profit or someone else's."

Haldane's contact with chemical warfare began in great disappointment. After the first German gas attack at Ypres, the British War Office, by Lord Kitchener's direct command, dispatched Haldane's father, the eminent respiratory physiologist John Scott Haldane, to France in a desperate effort to overcome this new danger. The elder Haldane, who had worked with his son on physiological experiments for many years, greatly valued both J.B.S.'s mathematical skills and his willingness to act as a human guinea pig in medical experiments (an ancient tradition among biologists and a favorite strategy of the elder Haldane, who never asked his son to do anything he wouldn't try on himself). So J.B.S., much to his initial disgust, left the front lines he loved so well and moved into a laboratory with his father.

J.B.S. already knew a great deal about toxic gases, primarily through his role as father's helper in self-experimentation. He recalled some early work with his father on firedamp (methane) in mines:

> To demonstrate the effects of breathing firedamp, my father told me to stand up and recite Mark Antony's speech from Shakespeare's *Julius Caesar,* beginning "Friends, Romans, countrymen." I soon began to pant, and somewhere about "the noble Brutus" my legs gave way and I collapsed on to the floor, where, of course, the air was all right. In this way I learnt that firedamp is lighter than air and not dangerous to breathe.

(Have you ever read a testimony more congruent with the stereotype of British upper-class intellectual dottiness?)

The Haldanes, *père et fils,* led a team of volunteer researchers in vitally important work (no doubt saving many thousands of lives) on the effects of noxious substances and the technology of gas masks. As always, they performed the most unpleasant and dangerous experiments on themselves. J.B.S. recalled:

> We had to compare the effects on ourselves of various quantities, with and without respirators. It stung the eyes and produced a tendency to gasp and cough when breathed. . . . As each of us got sufficiently affected by gas to render his lungs duly irritable, another would take his place. None of us was much the worse for the gas, or in any real danger, as we knew where to stop, but some had to

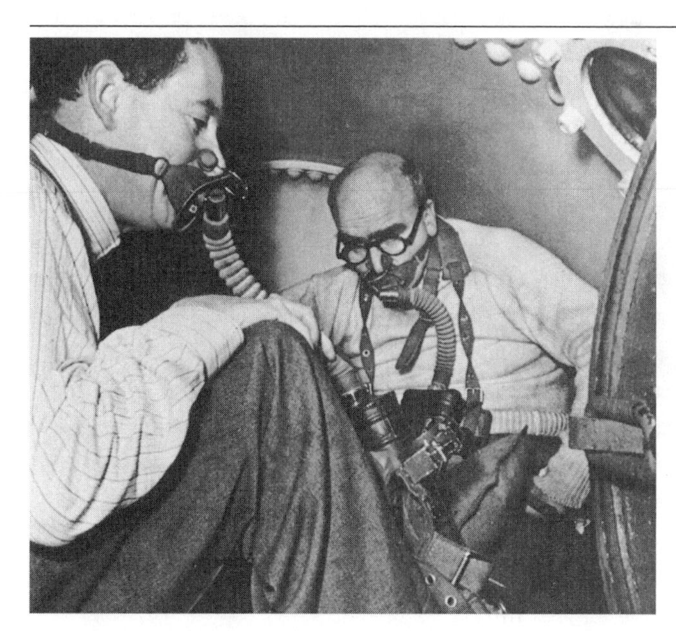

J.B.S. Haldane engaged in physiological self-experimentation on the effects of various gases.

> go to bed for a few days, and I was very short of breath and incapable of running for a month or so.

Thus, we cannot deny Haldane's superior knowledge or his maximal experience in the subject of chemical warfare. He therefore becomes an interesting test for the proposition that such expertise should confer special powers of forecasting—and that the technical knowledge of such people should therefore be trusted if they advocate a path of further development against the caution, the pessimism, even the defeatism of others who prefer moral restraint upon future technological progress because they fear the power of unforeseen directions and unanticipated consequences.

In 1925, as nations throughout the world signed the Geneva Protocol to ban chemical and biological warfare, J. B. S. Haldane published the most controversial of all his iconoclastic books: a slim volume of eighty-four pages entitled *Callinicus: A Defense of Chemical Warfare,* based on a lecture he had given in 1924. (Callinicus, a seventh-century Jewish refugee in Constantinople, invented Greek fire, an incendiary liquid that could be shot from siphons toward enemy ships or troops. The subsequent flames, almost impossible to extinguish, helped save the Byzantine empire from Islamic conquest for several centuries. The formula, known only to the emperor and to Callinicus' family, who held an exclusive right of manufacture, remained a state secret and still elicits controversy among scholars of warfare.)

Haldane's argument can be easily outlined. He summarized the data, including death tolls and casualty rates, from gas attacks in World War I and proclaimed the results more humane than the consequences of conventional weaponry.

> A case can be made out for gas as a weapon on humanitarian grounds, based on the very small proportion of killed to casualties from gas in the War, and especially during its last year [when better gas masks had been made and widely distributed].

Haldane based this conclusion on two arguments. He first listed the chemical agents used in the war and branded most of them as not dangerous for having only transient effects (making the assumption that temporarily insensate soldiers would be passed by or humanely captured rather than slaughtered). He regarded the few chemicals that could induce more permanent harm—mustard gas, in particular—as both hard to control and relatively easy to avoid, with proper equipment. Second, he called upon his own frequent experience with poison gases and stated a strong preference for these agents over his equally personal contact with bullets:

> Besides being wounded, I have been buried alive, and on several occasions in peacetime I have been asphyxiated to the point of unconsciousness. The pain and discomfort arising from the other experiences were utterly negligent compared with those produced by a good septic shell wound.

Haldane therefore concluded that gas, for reasons of effectiveness as a weapon and relative humaneness in causing few deaths compared with the number of temporary incapacitations, should be validated and further developed as a primary military tactic:

> I certainly share their [pacifists'] objection to war, but I doubt whether by objecting to it we are likely to avoid it in future, however lofty our motives or disinterested our conduct. . . . If we are to have more wars, I prefer that my country should be on the winning side. . . . If it is right for me to fight my enemy with a sword, it is right for me to fight him with mustard gas; if the one is wrong, so is the other.

I do not flinch before this last statement from the realm of ultimate realpolitik. The primary and obvious objection to Haldane's thesis in *Callinicus*—not only as raised now by me in the abstract, but also as advanced by Haldane's

World War I paraphernalia for protection from poison gas attacks.

numerous critics in 1925—holds that, whatever the impact of poison gas in its infancy in World War I (and I do not challenge Haldane's assessment), unrestrained use of this technology may lead to levels of effectiveness and numbers of deaths undreamed of in earlier warfare. Better the devil we know best than a devil seen only as an ineffective baby just introduced into our midst. If we can squelch this baby now, by moral restraint and international agreement, let's do so before he grows into a large and unstoppable adult potentially far more potent than any devil we know already.

(I should offer the proviso that, in making this general argument for moral restraint, I am speaking only of evident devils, or destructive technologies with no primary role in realms usually designated as human betterment: healing the sick, increasing agricultural yields, and so on. I am *not* talking about the more difficult, and common, problem of new technologies—cloning comes to mind as the current topic of greatest interest [see chapter 19]—with powerfully benevolent intended purposes but also some pretty scary potential misuses in the wrong hands, or in the decent hands of people who have not pondered the unintended consequences of good deeds. Such technologies may be regulated, but surely should not be banned.)

Haldane's response to this obvious objection reflects all the arrogance described in the first part of this essay: I have superior scientific knowledge of this subject and can therefore be trusted to forecast future potentials and dangers; from what I know of chemistry, and from what I have learned from the data of World War I, chemical weapons will remain both effective and relatively humane and should therefore be further developed. In other words, and in epitome: trust me.

> One of the grounds given for objection to science is that science is responsible for such horrors as those of the late War. "You scientific men (we are told) never think of the possible applications of your discoveries. You do not mind whether they are used to kill or to cure. Your method of thinking, doubtless satisfactory when dealing with molecules and atoms, renders you insensible to the difference between right and wrong." . . . The objection to scientific weapons such as the gases of the late War, and such new devices as may be employed in the next, is essentially an objection to the unknown. Fighting with lances or guns, one can calculate, or thinks one can calculate, one's chances. But with gas or rays or microbes one has an altogether different state of affairs.

> . . . What I have said about mustard gas might be applied, *mutatis mutandis,* to most other applications of science to human life. They can all, I think, be abused, but none perhaps is always evil; and many, like mustard gas, when we have got over our first not very rational objections to them, turn out to be, on the whole, good.

In fact, Haldane didn't even grant moral arguments—or the imposition of moral restraints—any role at all in the prevention of war. He adopted the same parochial and arrogant position, still all too common among scientists, that war can be ended only by rational and scientific research: "War will be prevented only by a scientific study of its causes, such as has prevented most epidemic diseases."

I am no philosopher, and I do not wish to combat Haldane's argument on theoretical grounds here. Let us look instead at the basic empirical evidence, unwittingly presented by Haldane himself in *Callinicus.* I therefore propose the following test: if Haldane's argument should prevail, and scientific recommendations should be trusted because scientists can forecast the future in areas of their expertise, then the success of Haldane's own predictions will validate his approach.

I propose that two great impediments generally stand in the way of successful prediction: first, our inability, in principle, to know much about complex futures along the contingent and nondeterministic pathways of history; and second, the personal hubris that leads us to think we act in a purely and abstractly rational manner, when our views really arise from unrecognized social and personal prejudices.

Callinicus contains an outstanding example of each error, and I rest my case for moral restraint here. Haldane does consider the argument that further development of chemical and biological weapons might prompt an investigation into even more powerful technologies of destruction—in particular, to unleashing the forces of the atom. But he dismisses this argument on scientific grounds of impossible achievement:

> Of course, if we could utilize the forces which we now know to exist inside the atom, we should have such capacities for destruction that I do not know of any agency other than divine intervention which would save humanity from complete and peremptory annihilation. . . . [But] we cannot utilize subatomic phenomena. . . . We cannot make apparatus small enough to disintegrate or fuse atomic nuclei. . . . We can only bombard them with particles of which

> perhaps one in a million hit, which is like firing keys at a safe-door from a machine gun a mile away in an attempt to open it. . . . We know very little about the structure of the atom and almost nothing about how to modify it. And the prospect of constructing such an apparatus seems to me to be so remote that, when some successor of mine is lecturing to a party spending a holiday on the moon, it will still be an unsolved (though not, I think, an ultimately unsolvable) problem.

To which, we need only reply: Hiroshima, 1945; Mr. Armstrong on the Moon, 1969. And we are still here—in an admittedly precarious atomic world—thanks to moral and political restraint.

But the even greater danger of arrogant and "rational" predictions unwittingly based on unrecognized prejudice led Haldane to the silliest statement he ever made—one that might be deemed socially vicious if our laughter did not induce a more generous mood. Haldane tries to forecast the revised style of warfare that mustard gas must impose upon future conflicts. He claims that some people have a natural immunity, differently distributed among our racial groups. He holds that 20 percent of whites, but 80 percent of blacks, are unaffected by the gas. Haldane then constructs a truly dotty scenario for future gas warfare: vanguards of black troops will lead the attack; German forces, with less access to this aspect of human diversity, might suffer some disadvantage, but their superior chemical knowledge should see them through, and balances should therefore be maintained:

> It seems, then, that mustard gas would enable an army to gain ground with far less killed on either side than the methods used in the late War, and would tend to establish a war of movement leading to a fairly rapid decision, as in the campaigns of the past. It would not upset the present balance of power, Germany's chemical industry being counterposed by French negro troops. Indians [that is, East Indians available to British forces] may be expected to be nearly as immune as negroes.

But now Haldane sees a hole in his argument. He steps back, breathes deeply, and finds a solution. Thank God for that 20 percent immunity among whites!

> The American Army authorities made a systematic examination of the susceptibility of large numbers of recruits. They found that there

> was a very resistant class, comprising 20% of the white men tried, but no less than 80% of the negroes. This is intelligible, as the symptoms of mustard gas blistering and sunburn are very similar, and negroes are pretty well immune to sunburn. It looks, therefore, as if, after a slight preliminary test, it should be possible to obtain colored troops who would all be resistant to mustard gas blistering in concentrations harmful to most white men. Enough resistant whites are available to officer them.

I am simply astonished (and also bemused) that this brilliant man, who preached the equality of humankind in numerous writings spanning more than fifty years, could have been so mired in conventional racial prejudices, and so wedded to the consequential and standard military practices of European and American armies, that he couldn't expand his horizons far enough even to imagine the possibility of competent black officers—and therefore had to sigh in relief at the availability of a few good men among the rarely resistant whites. If Haldane couldn't anticipate even this minor development in human relationships and potentialities, why should we trust his judgments about the far more problematical nature of future wars?

(This incident should carry the same message for current discussions about underrepresentation of minorities as managers of baseball teams or as quarterbacks in football. I also recall a famous and similar episode of ridiculously poor prediction in the history of biological determinism—the estimate by a major European car manufacturer, early in the century, that his business would be profitable but rather limited. European markets, he confidently predicted, would never require more than a million automobiles—for only so many men in the lower classes possessed sufficient innate intellectual ability to work as chauffeurs! Don't you love the triply unacknowledged bias of this statement—that poor folks rarely rank high in fixed genetic intelligence and that neither women nor rich folks could ever be expected to drive a car?)

The logic of my general argument must lead to a truly modest proposal. Wouldn't we all love to fix the world in one fell swoop of proactive genius? We must, of course, never stop dreaming and trying. But we must also temper our projects with a modesty born of understanding that we cannot predict the future and that the best-laid plans of mice and men often founder into a deep pit dug by unanticipated consequences. In this context, we should honor what might be called the "negative morality" of restraint and consideration, a principle that wise people have always understood (as embodied in the golden rule) and dreamers have generally rejected, sometimes for human good but more often for the evil that arises when demagogues and zealots try to impose their "true belief" upon all humanity, whatever the consequences.

The Hippocratic oath, often misunderstood as a great document about general moral principles in medicine, should be read as a manifesto for protecting the secret knowledge of a guild and for passing skills only to designated initiates. But the oath also includes a preeminent statement, later recast as a Latin motto for physicians, and ranking (in my judgment) with the Socratic dictum "know thyself" as one of the two greatest tidbits of advice from antiquity. I can imagine no nobler rule of morality than this single phrase, which every human being should engrave into heart and mind: *primum non nocere*—above all, do no harm.

VI

Evolution at All Scales

21

Of Embryos and Ancestors

"EVERY DAY, IN EVERY WAY, I'M GETTING BETTER and better." I had always regarded this famous phrase as a primary example of the intellectual vacuity that often passes for profundity in our current era of laid-back, New Age bliss—a verbal counterpart to the vapidity of the "have a nice day" smiley face. But when I saw this phrase chiseled in stone on the pediment of a French hospital built in the early years of our century, I knew that I must have missed a longer and more interesting pedigree. This formula for well-being, I then discovered, had been devised in 1920 by Emile Coué (1857–1926), a French pharmacist who made quite a stir in the pop-psych circles of his day with a theory of self-improvement through autosuggestion based on frequent repetition of this mantra—a treatment that received the name of Couéism. (In a rare example of improvement in translation, this phrase gains both a rhyme and better flow, at least to my ears, when converted to English from Coué's French original—*"tous les jours, à tous les points de vue, je vais de mieux en mieux."*)

I don't doubt the efficacy of Coué's mantra, for the "placebo effect" (its only possible mode of action) should not be dismissed as a delusion, but cherished as a useful strategy for certain forms of healing—a primary example of the influence that mental attitudes can wield upon our physical sense of well-being. However, as a general description for the usual style and pacing of human improvement, the constant and steady incrementalism of Coué's motto—a twentieth-century version of an ancient claim embodied in the victory cry of Aesop's tortoise, "slow and steady wins the race"—strikes me as only rarely applicable, and surely secondary to the usual mode of human enlightenment, either attitudinal or intellectual: that is, *not* by global creep forward, inch by subsequent inch, but rather in rushes or whooshes, usually following the removal of some impediment, or the discovery of some facilitating device, either ideological or technological.

The glory of science lies in such innovatory bursts. Centuries of vain speculation dissolved in months before the resolving power of Galileo's telescope, trained upon the full range of cosmic distances, from the moon to the Milky Way (see chapter 2). About 350 years later, centuries of conjecture and indirect data about the composition of lunar rocks melted before a few pounds of actual samples, brought back by Apollo after Mr. Armstrong's small step onto a new world.

In the physical sciences, such explosions of discovery usually follow the invention of a device that can, for the first time, penetrate a previously invisible realm—the "too far" by the telescope, the "too small" by the microscope, the imperceptible by X-rays, or the unreachable by spaceships. In the humbler world of natural history, episodes of equal pith and moment often follow a eureka triggered by continually available mental, rather than expensively novel physical, equipment. In other words, great discovery often requires a map to a hidden mine filled with gems then easily gathered by conventional tools, not a shiny new space-age machine for penetrating previously inaccessible worlds.

The uncovering of life's early history has featured several such cascades of discovery following a key insight about proper places to look—and I introduce this year's wonderful story by citing a previous incident of remarkably similar character from the last generation of our science (literally so, for this year's discoverer wrote his Ph.D. dissertation under the guidance of the first innovator).

When, as a boy in the early 1950s, I first became fascinated with paleontology and evolution, the standard dogma about the origin of life proclaimed such an event inherently improbable, but achieved on this planet only because the immensity of geological time must convert the nearly impossible into the virtually certain. (With no limit on the number of tries, you will eventually flip fifty heads in a row with an honest coin.) As evidence for asserting the exquisite specialness of life in the face of overwhelmingly contrary odds, these conventional sources cited the absence of any fossils representing the first half of the earth's existence—a span of more than two billion years, often formally designated on older geological charts as the Azoic (literally "lifeless") era. Although scientists do recognize the limitations of such "negative evidence" (the first example of a previously absent phenomenon may, after all, turn up tomorrow), this failure to find any fossils for geology's first two billion years did seem fairly persuasive. Paleontologists had been searching assiduously for more than a century and had found nothing but ambiguous scraps and blobs. Negative results based on such sustained effort over so many years do begin to inspire belief.

But the impasse broke in the 1950s, when Elso Barghoorn and Stanley Tyler reported fossils of unicellular life in rocks more than two billion years old. Paleontologists, to summarize a long and complex story with many exciting turns and notable heroes, had been looking in the wrong place—in conventional sediments that rarely preserve the remains of single-celled bacterial organisms without hard parts. We had not realized that life had remained so simple for so long, or that the ordinary sites for good fossil records could not preserve such organisms.

Barghoorn and colleagues dispelled a century of frustration by looking in a different place, where cellular remains of bacteria might be preserved—in chert beds. Chert has the same chemical formula (with a different molecular arrangement) as quartz: silicon dioxide. Paleontologists rarely think of looking for fossils in silicate rocks—for the perfectly valid and utterly obvious reason that silicates form by the cooling of volcanic magmas and therefore cannot contain organic remains. (Life, after all, doesn't flourish in bubbling lavas, and anything falling in gets burnt to a crisp.) But cherts can form at lower temperatures and be deposited amid layers of ordinary sediments in oceanic waters. Bacterial cells, when trapped in this equivalent of surrounding glass, can be preserved as fossils.

This cardinal insight—that we had been searching in the wrong venue of ordinary (and barren) sediments rather than in fruitful cherts—created an entire field of study: collecting data for the first two-thirds or so of life's full history! Forty years later, we may look back with wonder at the flood of achievement and the complete overturn of established wisdom. We now possess a rich fossil record of early life, extending right back to the earliest potential source for cellular evidence. (The oldest rocks on earth that could preserve such data do contain abundant fossils of bacterial organisms. These 3.5-to-3.6-billion-year-old rocks from Australia and South Africa are the most ancient strata on earth that have not been sufficiently altered by subsequent heat and pressure to destroy all anatomical evidence of life.)

Such ubiquity and abundance has forced a reversal of the old view. The origin of life at simplest bacterial grade now seems inevitable rather than improbable. As a mantra for memory, may I suggest: "Life on earth is as old as it could be." I realize, of course, that an earliest possible appearance builds no proof for inevitability. After all, even a highly improbable event might occur, by good fortune, early in a series of trials. (You might flip those fifty successive heads on your tenth attempt—but don't count, or bet, on it!) Nonetheless, faced with the data we now possess—that life appeared as soon as environmental conditions permitted such an event and then remained pervasive forever after—our thoughts must move to ideas about almost predictive inevitability. Given a planet of earthly size, distance, and composition, life of simplest grade probably originates with virtual certainty as a consequence of principles of organic chemistry and the physics of self-organizing systems.

But whatever the predictability of life's origin, subsequent pathways of evolution have run in mighty peculiar directions, at least with respect to our conventional hopes and biases. The broadest pattern might seem to confirm our usual view of generally increasing complexity, leading sensibly to human consciousness: after all, the early earth sported only bacteria, while our planet now boasts people, ant colonies, and oak trees. Fair enough, but closer scrutiny of general timings or particular details can promote little faith in any steady pattern. If greater size and complexity provide such Darwinian blessings, why did life take so long to proceed "onward," and why do most of the supposed steps occur so quirkily and so quickly? Consider the following epitome of major events:

Fossils, as stated above, appear as soon as they possibly could in the geological record. But life then remains exclusively at this simplest so-called prokaryotic grade (unicells without any internal organelles, that is, no nuclei, chromosomes, mitochondria, and so on) for about half its subsequent history. The first unicells of the more complex eukaryotic grade (with the conventional organelles of our high school textbook pictures of an amoeba or paramecium) do not appear in the fossil record until about two billion years ago. The three great multicellular kingdoms of plants, fungi, and animals arise subsequently (and, at least for algae within the plant kingdom, more than once and independently) from eukaryotic unicells. Fossils of simple multicellular algae extend back fairly reliably to about one billion years, and far more conjecturally to as many as 1.8 billion years.

But the real enigma—at least with respect to our parochial concerns about the progressive inevitability of our own lineage—surrounds the origin and early history of animals. If life had always been hankering to reach a pinnacle of expression as the animal kingdom, then organic history seemed in no hurry to initiate this ultimate phase. About five-sixths of life's time had elapsed before animals made their first appearance in the fossil record some 600 million years ago. Moreover, the earth's first community of animals, which held nearly exclusive sway from an initial appearance some 600 million years ago right to the dawn of the Cambrian period 543 million years ago, consisted of enigmatic species with no clear relation to modern forms.

These so-called Ediacaran animals (named for the locality of first discovery in Australia, but now known from all continents) could grow quite large (up to a few feet in length), but apparently contained neither complex internal organs, nor even any recognizable body openings for mouth or anus. Many Ediacaran creatures were flattened, pancakelike forms (in a variety of shapes and sizes), built of numerous tubelike sections, complexly quilted together into a single structure. Theories about the affinities of Ediacaran organisms span the full gamut—from viewing them, most conventionally, as simple ancestors for several modern phyla; to interpreting them, most radically, as an entirely separate (and ultimately failed) experiment in multicellular animal life. An intermediate position, now gaining favor—a situation that should lead to no predictions about the ultimate outcome of this complex debate—treats Ediacaran animals as a bountiful expression of the range of possibilities for diploblastic animals (built of two body layers), a group now so reduced in diversity (and subsisting only as corals, jellyfishes, and their allies) that living representatives provide little understanding of full potentials.

Modern animals—except for sponges, corals, and a few other minor groups—are all triploblastic, or composed of three body layers: an ectoderm, forming nervous tissue and other organs; mesoderm, forming reproductive structures and other parts; and endoderm, building the gut and other internal organs. (If you learned a conventional list of phyla back in high school biology, all groups from the flatworms on "up," including all the "big" phyla of annelids, arthropods, mollusks, echinoderms, and vertebrates, are triploblasts.) This three-layered organization seems to act as a prerequisite for the formation of conventional, complex, mobile, bilaterally symmetrical organisms with body cavities, appendages, sensory organs, and all the other accoutrements that set our standard picture of a "proper" animal. Thus, in our aimlessly parochial manner (and ignoring such truly important groups as corals and sponges), we tend to equate the problem of the beginning of modern animals with the origin of triploblasts. If the Ediacaran animals are all (or mostly) diploblasts, or even more genealogically divergent from triploblast animals, then this first fauna does not resolve the problem of the origin of animals (in our conventionally limited sense of modern triploblasts).

The story of modern animals then becomes even more curious. The inception of the Cambrian period, 543 million years ago, marks the extinction, perhaps quite rapidly, of the Ediacara fauna, and the beginning of a rich record for animals with calcareous skeletons easily preserved as fossils. But the first phase of the Cambrian, called Manakayan and lasting from 543 to 530 million years ago, primarily features a confusing set of spines, plates, and other bits and pieces called (even in our technical literature) the SSF, or "small shelly fossils" (presumably the disarticulated fragments of skeletons that had not yet evolved to large, discrete units covering the entire organism).

The next two phases of the Cambrian (called Tommotian and Atdabanian, and ranging from 530 to about 520 million years ago) mark the strangest, most important, and most intriguing of all episodes in the fossil record of animals—the short interval known as the "Cambrian explosion," and featuring the first fossils of all animal phyla with skeletons subject to easy preservation in the fossil record. (A single exception, a group of colonial marine organisms called the

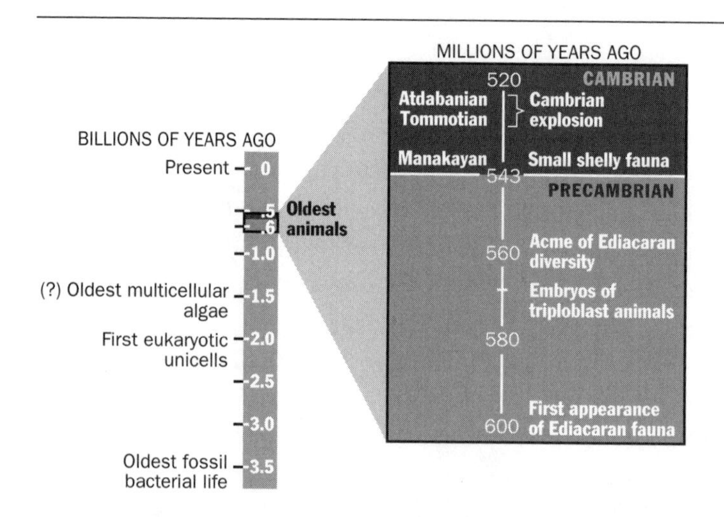

Time charts for major events in the history of life (left) and for details of the Cambrian Explosion and other events in the origin of multicellular animals (right).

Haeckel's theoretical drawings of ancestral animals (left) compared with fossils of a Precambrian embryo (below).

Bryozoa, make their appearance at the beginning of the next or Ordovician period. Many intriguing inventions, including human consciousness and the dance language of bees, have arisen since then, but no new phyla, or animals of starkly divergent anatomical design.)

The Cambrian explosion ranks as such a definitive episode in the history of animals that we cannot possibly grasp the basic tale of our own kingdom until we achieve better resolution for both the antecedents and the unfolding of this cardinal geological moment. A major discovery, announced in February 1998, and also based on learning to look in a previously unsuspected place, has thrilled the entire paleontological community for its promise in unraveling the previously unknown history of triploblast animals before the Cambrian explosion.

If the Cambrian explosion inspires frustration for its plethora of data—too much, too confusing, and too fast—the Precambrian history of triploblast animals engenders even more chagrin for its dearth. The complex animals of the explosion, so clearly assignable to modern phyla, obviously didn't arise *ex nihilo* at their first moment of fossilization, but who (and where) are their antecedents in Precambrian times? What were the forebears of modern animals doing for 50 million prior years, when Ediacaran diploblasts (or stranger creatures) ruled the animal world?

Up to now, we have engaged in much speculation, while possessing only a whiff or two of data. Ediacaran strata also contain trails and feeding traces presumably made by triploblast organisms of modern design (for the flattened and mostly immobile Ediacaran animals could not crawl, burrow, or feed in a manner so suggestive of activities now confined to triploblast organisms). Thus, we do have evidence for the existence, and even the activities, of precursors for modern animals before the Cambrian explosion, but no data at all about their anatomy and appearance—a situation akin to the frustration we might feel if we could hear birdsong but had never seen a bird.

A potential solution—or at the very least, a firm and first source of anatomical data—has just been discovered by applying the venerable motto (so beloved by people, including yours truly, of shorter-than-average stature): good things often come in small packages, or to choose a more literary and inspirational expression, Micah's statement (5:2) taken by the later evangelists as a prophecy of things to come: "But thou, Bethlehem . . . though thou be little among the thousands of Judah, yet out of thee shall he come forth unto me that is to be ruler in Israel."

In short, paleontologists had been looking for conventional fossils in the usual (and visible) size ranges of adult organisms: fractions to few inches. But a solution had been lurking at the smaller size of creatures just barely visible (in principle) but undetectable in conventional practice—in the domain of embryos. Who would ever have thought that delicate embryos might be preserved as fossils, when presumably hardier adults left no fragments of their existence? The story, a fascinating lesson in the ways of science, has been building for more than a decade, but has only just been extended to the problem of Precambrian animals.

Fossils form in many modes and styles—as original hard parts preserved within entombing sediments, or as secondary structures formed by impressions of bones or shells (molds) that may then become filled with later sediments (casts). But original organic materials may also be replaced by percolating minerals—a process called petrifaction, or literally "making into stone," a phenomenon perhaps best represented in popular knowledge by gorgeous specimens from the Petrified Forest of Arizona, where multicolored agate (another form of silicon dioxide) has replaced original carbon so precisely that the wood's cellular structure can still be discerned. (Petrifaction enjoys sufficient public renown that many people mistakenly regard such replacement as the primary definition of a fossil. But any bit of an ancient organism qualifies as a fossil, whatever the style of preservation. In almost any circumstance, a professional would much prefer to work with unaltered hard parts than with petrified replacements.)

In any case, one poorly understood style of petrifaction leads to replacement of soft tissues by calcium phosphate—a process called phosphatization. This style of replacement can occur within days of death, thus leading to the rare and precious phenomenon of petrifaction before decay of soft anatomy. Phosphatization might provide a paleontologist's holy grail if all soft tissues could thus be preserved at any size in any kind of sediment. Alas, the process seems to work in detail only for tiny objects up to about two millimeters in length (26.4 millimeters make an inch, so we are talking about barely visible dots, not even about bugs large enough to be designated as "yucky" when found in our dinner plates or beds).

Still, on the good old principle of not looking gift horses (or unexpected bounties) in the mouth (by complaining about an unavailable better deal), let us rejoice in the utterly unanticipated prospect that tiny creatures—which are, after all, ever so abundant in nature, however much they may generally pass beneath our exalted notice—might become petrified in sufficient detail to preserve their bristles, hairs, or even their cellular structure. The recognition that phosphatization may open up an entire world of tiny creatures, previously never considered as candidates for fossilization at all, may spark the greatest burst of paleontological exploration since the discovery that two billion years of Precambrian life lay hidden in chert.

The first hints that exquisite phosphatization of tiny creatures might resolve key issues in the early evolution of animals date to a discovery made in the mid-1970s and then researched and reported in one of the most elegant, but rather sadly underappreciated, series of papers ever published in the history of paleontology: the work of two German scientists, Klaus J. Müller and Dieter Walossek, on the fauna of distinctive upper Cambrian rocks in Sweden, known as Orsten beds. In these layers of limestone concretions, tiny arthropods (mostly larvae of crustaceans) have been preserved by phosphatization in exquisite, three-dimensional detail. The photography and drawings of Walossek and Müller have rarely been equaled in clarity and aesthetic brilliance, and their papers are a delight both to read and to see. (For a good early summary, consult K. J. Müller and D. Walossek, "A remarkable arthropod fauna from the Upper Cambrian 'Orsten' of Sweden," *Transactions of the Royal Society of Edinburgh,* 1985, volume 76, pages 161–72; for a recent review, see Walossek and Müller, "Cambrian 'Orsten'-type arthropods and the phylogeny of Crustacea," in R. A. Fortey and R. H. Thomas, eds., *Arthropod Relationships,* London: Chapman and Hall, 1997.)

By dissolving the limestone in acetic acid, Walossek and Müller can recover the tiny phosphatized arthropods intact. They have collected more than one hundred thousand specimens following this procedure and have summarized their findings in a recent paper of 1997:

> The cuticular surface of these arthropods is still present in full detail, revealing eyes and limbs, hairs and minute bristles, . . . gland openings, and even cellular patterns and grooves of muscle attachments underneath. . . . The maximum size of specimens recovered in this type of preservation does not exceed 2 mm.

From this beginning, other paleontologists have proceeded backward in time, and downward in growth from larvae to early embryonic stages containing just a few cells. In 1994, Xi-guang Zhang and Brian R. Pratt found balls of presumably embryonic cells measuring 0.30 to 0.35 millimeter in length and representing, perhaps, the earliest stages of adult trilobites also found in the same Middle Cambrian strata (Zhang and Pratt, "Middle Cambrian arthropod embryos with blastomeres," *Science,* 1994, volume 266, pages 637–38). In 1997, Stefan Bengston and Yue Zhao then reported even earlier phosphatized embryos from basal Cambrian strata in China and Siberia. In an exciting addition to this growing literature, these authors traced a probable growth series, from embryos to tiny near adults, for two entirely different animals: a species from an enigmatic extinct group, the conulariids; and a probable segmented worm (Bengston and Zhao, "Fossilized metazoan embryos from the earliest Cambrian," *Science,* 1997, volume 277, pages 1645–48).

When such novel technologies first encounter materials from a truly unknown or unsuspected world, genuinely revolutionary conclusions often emerge. In what may well be regarded by subsequent historians as the greatest paleontological discovery of the late twentieth century, Shuhai Xiao, a postdoctoral student in our paleontological program, Yun Zhang of Beijing University, and my colleague, and Shuhai Xiao's mentor, Andrew H. Knoll, have just reported their discovery of the oldest triploblastic animals, preserved as phosphatized embryos in rocks from southern China estimated at 570 million years in age—and thus even older than the best-preserved Ediacaran faunas, found in strata about 10 million years younger (see Xiao, Zhang, and Knoll, "Three-dimensional preservation of algae and animal embryos in a Neoproterozoic phosphorite," *Nature,* 1998, volume 391, pages 553–58). These phosphatized fossils include a rich variety of multicellular algae, showing, according to the authors, that "by the time large animals enter the fossil record, the three principal groups of multicellular algae had not only diverged from other protistan [unicellular] stocks but had evolved a surprising degree of the morphological complexity exhibited by living algae."

Still, given our understandably greater interest in our own animal kingdom, most attention will be riveted upon some smaller and rarer globular fossils, averaging half a millimeter in length, and found phosphatized in the same strata: an exquisite series of earliest embryonic stages, beginning with a single fertilized egg and proceeding through two-cell, four-cell, eight-cell, and sixteen-cell stages to small balls of cells representing slightly later phases of early development. These embryos cannot be assigned to any particular group (as more distinctive later stages have not yet been found), but their identification as earliest stages of triploblastic animals seems secure, both from characteristic features (especially the unchanging overall size of the embryo during these earliest stages, as average cell size decreases to pack more cells into a constant space), and uncanny resemblance to particular traits of living groups. (Several embryologists have told Knoll and colleagues that they would have identified these specimens as embryos of living crustaceans had they not been informed of their truly ancient age!)

Elso Barghoorn, Knoll's thesis adviser, opened up the world of earliest life by discovering that bacteria could be preserved in chert. Now, a full generation later, Knoll and colleagues have penetrated the realm of earliest known animals of modern design by accessing a new domain where phosphatization preserves minute embryonic stages, but no known process of fossilization can reliably render potentially larger phases of growth. When I consider the cascade of knowledge that proceeded from Barghoorn's first report of Precambrian bacteria to our current record spanning three billion Precambrian years and hundreds of recorded forms, I can only conclude that the discovery of Xiao, Zhang, and Knoll places us at a gateway of equal promise for reconstructing the earliest history of modern animals, before their overt evolutionary burst to large size and greatly increased anatomical variety in the subsequent Cambrian explosion. If we can thereby gain any insight into the greatest of all mysteries surrounding the early evolution of animals—the causes of both the anatomical explosion itself and the "turning off" of evolutionary fecundity for generating new phyla thereafter—then paleontology will shake hands with evolutionary theory in the finest merger of talents ever applied to the resolution of a historical enigma.

A closing and more general commentary may help to set a context of both humility and excitement at the threshold of this new quest. First, we might be able to coordinate the direct evidence of fossils with a potentially powerful indirect method for judging the times of origin and branching for major animal groups: the measurement of relative degrees of detailed genetic similarity among living representatives of diverse animal phyla. Such measurements can be made with great precision upon large masses of data, but firm conclusions do not always follow because various genes evolve at different rates that also maintain no constancy over time—and most methods applied so far have made simplifying (and probably unjustified) assumptions about relatively even ticking of supposed molecular clocks.

For example, in a paper that received much attention upon publication in 1996, G. A. Wray, J. S. Levinton, and L. H. Shapiro used differences in the molecular sequences of seven genes in living representatives of major phyla to derive an estimate of roughly 1.2 billion years for the divergence time between chordates (our phylum) and the three great groups on the other major genealogical branch of animals (arthropods, annelids, and mollusks), and 1.0 billion years for the later divergence of chordates from the more closely related phylum of echinoderms (Wray, Levinton, and Shapiro, "Molecular evidence for deep Precambrian divergences among metazoan phyla," *Science,* 1996, volume 274, pages 568–73).

This paper sowed a great deal of unnecessary confusion when several uncomprehending journalistic reports, and a few careless statements by the authors, raised the old and false canard that such an early branching time for animal phyla disproves the reality of the Cambrian explosion by rendering this apparent burst of diversity as the artifact of an imperfect fossil record (signifying, perhaps, only the invention of hard parts, rather than any acceleration of anatomical innovation). For example, Wray et al. write: "Our results cast doubt on the prevailing notion that the animal phyla diverged explosively during the Cambrian or late Vendian [Ediacaran times], and instead suggest that there was an extended period of divergence . . . commencing about a billion years ago."

But such statements confuse the vital distinction, in both evolutionary theory and actual results, between times of initial branching and subsequent rates of anatomical innovation or evolutionary change in general. Even the most vociferous advocates of a genuine Cambrian explosion have never argued that this period of rapid anatomical diversification marks the moment of origin for animal phyla—if only because we all acknowledged the evidence for Precambrian tracks and trails of triploblasts even before the recent discovery of embryos. Nor do these same vociferous advocates imagine that only one wormlike species crawled across the great Cambrian divide to serve as an immediate common ancestor for all modern phyla. In fact, I can't imagine why anyone would care (for adjudicating the reality of the explosion, though one would care a great deal for discussions of some other evolutionary issues) whether one wormlike species carrying the ancestry of all later animals, or ten similar wormlike species already representing the lineages of ten subsequent phyla, crossed this great divide from an earlier Precambrian history. The Cambrian explosion represents a claim for a rapid spurt of *anatomical innovation* within the animal kingdom, not an argument about times of *genealogical divergence.*

The following example should clarify the fundamental distinction between times of genealogical splitting and rates of change. Both rhinoceroses and horses may have evolved from the genus *Hyracotherium* (formerly called *Eohippus*). A visitor to the Eocene earth about 50 million years ago might determine that the basic split had already occurred. He might be able to identify one species of *Hyracotherium* as the ancestor of all later horses, and another species of the same genus as the progenitor of all subsequent rhinos. But this visitor would be laughed to justified scorn if he then argued that later divergences between horses and rhinos must be illusory because the two lineages had already split. After all, the two Eocene species looked like kissing cousins (as evidenced by their placement in the same genus), and only gained their later status as progenitors of highly distinct lineages by virtue of a subsequent history, utterly unknowable at the time of splitting. Similarly, if ten nearly identical wormlike forms (the analogs of the two *Hyracotherium* species) crossed the Cambrian boundary, but only evolved the anatomical distinctions of great phyla during the subsequent explosion, then the explosion remains as real, and as vitally important for life's history, as any advocate has ever averred.

This crucial distinction has been recognized by most commentators on the work of Wray et al. Geerat J. Vermeij, in his direct evaluation (*Science,* 1996, page 526), wrote that "this new work in no way diminishes the significance of the Vendian-Cambrian revolution." Fortey, Briggs, and Wills added (*BioEssays,* 1997, page 433) that "there is, of course, no necessary correspondence between morphology and genomic change." In any case, a later publication by Ayala, Rzhetsky, and Ayala (*Proceedings of the National Academy of Sciences,* 1998,

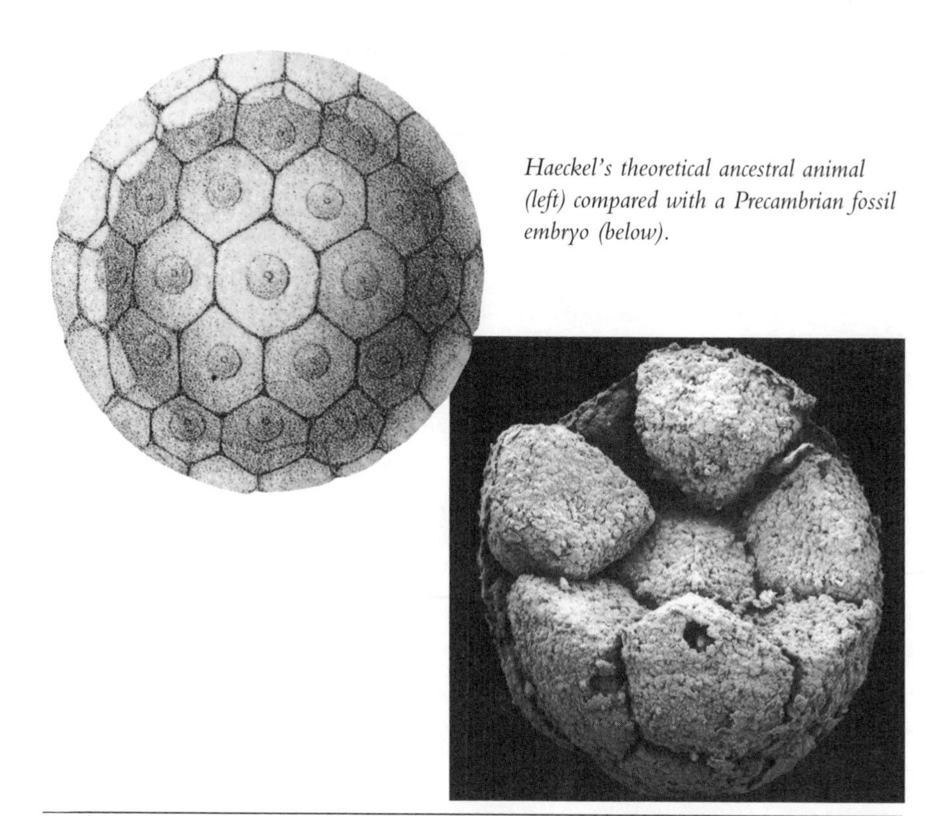

Haeckel's theoretical ancestral animal (left) compared with a Precambrian fossil embryo (below).

volume 95, pages 606–11) presents a powerful rebuttal to Wray et al.'s specific conclusions. By correcting statistical errors and unwarranted assumptions, and by adding data for twelve additional genes, these authors provide a very different estimate for initial diversification in late Precambrian times: about 670 million years ago for the split of chordates from the line of arthropods, annelids, and mollusks; and 600 million years for the later divergence of chordates from echinoderms.

We are left, of course, with a key mystery (among many others): where are Precambrian adult triploblasts "hiding," now that we have discovered their embryos? An old suggestion, first advanced in the 1870s in the prolific and often highly speculative work of the German biologist Ernst Haeckel (who was nonetheless outstandingly right far more often than random guesswork would allow), held that Precambrian animals had evolved as tiny forms not much larger than, or very different from, modern embryos—and would therefore be very hard to find as fossils. (The similarity between Haeckel's hypothetical ancestors and Xiao, Zhang, and Knoll's actual embryos is almost eerie—see figures on pages 323 and 330.) Moreover, E. H. Davidson, K. J. Peterson, and R. A. Cameron (*Science,* 1995, volume 270, pages 1319–25) have made a powerful case, based on genetic and developmental arguments, that Precambrian animals did originate at tiny sizes, and that the subsequent Cambrian explosion featured the evolution of novel embryological mechanisms for substantially increasing cell number and body size, accompanied by consequent potential for greatly enhanced anatomical innovation. If Haeckel's old argument, buttressed by Davidson's new concepts and data, has validity, we then gain genuine hope, even realistic expectation, that Precambrian adult triploblasts may soon be discovered, for such animals will be small enough to be preserved by phosphatization.

As a final point, this developing scenario for the early history of animals might foster humility and generate respect for the complexity of evolutionary pathways. To make the obvious analogy: we used to regard the triumph of "superior" mammals over "antediluvian" dinosaurs as an inevitable consequence of progressive evolution. We now realize that mammals originated at the same time as dinosaurs and then lived for more than 100 million years as marginal, small-bodied creatures in the nooks and crannies of a dinosaur's world. Moreover, mammals would never have expanded to dominate terrestrial ecosystems (and humans would surely never have evolved) without the supreme good fortune (for us) of a catastrophic extraterrestrial impact that, for some set of unknown reasons, eliminated dinosaurs and gave mammals an unanticipated opportunity.

Does this essay's tale from an earlier time—Ediacaran "primitives" versus contemporary Precambrian ancestors of modern animals—differ in any substantial way? We now know (from the evidence of Xiao, Zhang, and Knoll's embryos) that animals of modern design had already originated before the Ediacara fauna evolved into full bloom. Yet "primitive" Ediacara dominated the world of animal life, perhaps for 100 million years, while modern triploblasts waited in the proverbial wings, perhaps as tiny animals of embryonic size, living in nooks and crannies left over by much larger Ediacaran dominants. Only a mass extinction of unknown cause, a great dying that wiped out Ediacara and initiated the Cambrian transition 543 million years ago, gave modern triploblasts an opportunity to shine—and so we have.

In evolution, as well as in politics, incumbency offers such powerful advantages that even a putatively more competent group may be forced into a long period of watchful waiting, always hoping that an external stroke of good luck will finally grant an opportunity for picking up the reins of power. If fortune continues to smile, the new regime may eventually gain enough confidence to invent a comforting and commanding mythology about the inevitability of its necessary rise to dominance by gradually growing better and better—every day and in every way.

22

The Paradox of the Visibly Irrelevant

An odd principle of human psychology, well known and exploited by the full panoply of prevaricators, from charming barkers like Barnum to evil demagogues like Goebbels, holds that even the silliest of lies can win credibility by constant repetition. In current American parlance, these proclamations of "truth" by Xeroxing fall into the fascinating domain of "urban legends."

My favorite bit of nonsense in this category intrudes upon me daily, and in very large type, thanks to a current billboard ad campaign by a company that will remain nameless. The latest version proclaims: "Scientists say we use 10 percent of our brains. That's way too much." Just about everyone regards the "truth" of this proclamation as obvious and incontrovertible—though you might still start a barroom fight over whether the correct figure should be 10,

15, or 20 percent. (I have heard all three asserted with utter confidence.) But this particular legend can only be judged as even worse than false: for the statement is truly meaningless and nonsensical. What do we mean by "90 percent unused"? What is all this superfluous tissue doing? The claim, in any case, can have no meaning until we develop an adequate theory about how the brain works. For now, we don't even have a satisfactory account for the neurological basis of memory and its storage—surely the sine qua non for formulating any sensible notion about unused percentages of brain matter! (I think that the legend developed because we rightly sense that we ought to be behaving with far more intelligence than we seem willing to muster—and the pseudoquantification of the urban legend acts as a falsely rigorous version of this legitimate but vague feeling.)

In my field of evolutionary biology, the most prominent urban legend—another "truth" known by "everyone"—holds that evolution may well be the way of the world, but one has to accept the idea with a dose of faith because the process occurs far too slowly to yield any observable result in a human lifetime. Thus, we can document evolution from the fossil record and infer the process from the taxonomic relationships of living species, but we cannot see evolution on human timescales "in the wild."

In fairness, we professionals must shoulder some blame for this utterly false impression about evolution's invisibility in the here and now of everyday human life. Darwin himself—though he knew and emphasized many cases of substantial change in human time (including the development of breeds in his beloved pigeons)—tended to wax eloquent about the inexorable and stately slowness of natural evolution. In a famous passage from *The Origin of Species,* he even devised a striking metaphor about clocks to underscore the usual invisibility:

> It may be said that natural selection is daily and hourly scrutinizing, throughout the world, every variation, even the slightest; rejecting that which is bad, preserving and adding up all that is good; silently and invisibly working. . . . We see nothing of these slow changes in progress until the hand of time has marked the long lapse of ages.

Nonetheless, the claim that evolution must be too slow to see can only rank as an urban legend—though not a completely harmless tale in this case, for our creationist incubi can then use the fallacy as an argument against evolution at any scale, and many folks take them seriously because they just "know" that evolution can never be seen in the immediate here and now. In fact, a precisely opposite situation actually prevails: biologists have documented a veritable glut of cases for rapid and eminently measurable evolution on timescales of years and decades.

However, this plethora of documents—while important for itself, and surely valid as a general confirmation for the proposition that organisms evolve—teaches us rather little about rates and patterns of evolution at the geological scales that build the history and taxonomic structure of life. The situation is wonderfully ironic—a point that I have tried to capture in the title of this article. The urban legend holds that evolution is too slow to document in palpable human lifetimes. The opposite truth has affirmed innumerable cases of measurable evolution at this minimal scale—but to be visible at all over so short a span, evolution must be far too rapid (and transient) to serve as the basis for major transformations in geological time. Hence the "paradox of the visibly irrelevant"—or, "if you can see it at all, it's too fast to matter in the long run!"

Our best and most numerous cases have been documented for the dominant and most evolutionarily active organisms on our planet—bacteria. In the most impressive of recent examples, Richard E. Lenski and Michael Travisano (*Proceedings of the National Academy of Sciences,* 1994, volume 91, pages 6808–14) monitored evolutionary change for ten thousand generations in twelve laboratory populations of the common human gut bacterium *Escherichia coli.* By placing all twelve populations in identical environments, they could study evolution under ideal experimental conditions of replication—a rarity for the complex and unique events of evolutionary transformation in nature. In a fascinating set of results, they found that each population reacted and changed differently, even within an environment made as identical as human observers know how to do. Yet Lenski and Travisano did observe some important and repeated patterns within the diversity. For example, each population increased rapidly in average cell size for the first two thousand generations or so, but then remained nearly stable for the last five thousand generations.

A cynic might still reply: fine, I'll grant you substantial observable evolution in the frenzied little world of bacteria, where enormous populations and new generations every hour allow you to monitor ten thousand episodes of natural selection in a manageable time. But a similar "experiment" would consume thousands of years for multicellular organisms that measure generations in years or decades rather than minutes or hours. So we may still maintain that evolution cannot be observed in the big, fat, furry, sexually reproducing organisms that serve as the prototype for "life" in ordinary human consciousness. (A reverse cynic would then rereply that bacteria truly dominate life, and that vertebrates only represent a latecoming side issue in the full story of evolution, however falsely promoted to centrality by our own parochial focus. But we must leave this deep issue to another time.)

I dedicate this essay to illustrating our cynic's error. Bacteria may provide our best and most consistent cases for obvious reasons, but measurable (and substantial) evolution has also, and often, been documented in vertebrates and other complex multicellular organisms. The classic cases have not exactly been hiding their light under a bushel, so I do wonder why the urban legend of evolution's invisibility persists with such strength. Perhaps the firmest and most elegant examples involve a group of organisms named to commemorate our standard-bearer himself—Darwin's finches of the Galápagos Islands, where my colleagues Peter and Rosemary Grant have spent many years documenting fine-scale evolution in such adaptively important features as size and strength of the bill (a key to the mechanics of feeding), as rapid climatic changes force an alteration of food preferences. This work formed the basis for Jonathan Weiner's excellent book, *The Beak of the Finch*—so the story has certainly been well and prominently reported in both the technical and popular press.

Nonetheless, new cases of such short-term evolution still maintain enormous and surprising power to attract public attention—for interesting and instructive but utterly invalid reasons, as I shall show. I devote this essay to the three most prominent examples of recent publications that received widespread attention in the popular press as well. (One derives from my own research, so at least I can't be accused of sour grapes in the debunking that will follow—though I trust that readers will also grasp the highly positive twist that I will ultimately impose upon my criticisms.) I shall briefly describe each case, then present my two general critiques of their prominent reporting by the popular press, and finally explain why such cases teach us so little about evolution in the large, yet remain so important for themselves, and at their own equally legitimate scale.

1. Guppies from Trinidad. In many drainage systems on the island of Trinidad, populations of guppies live in downstream pools, where several species of fish can feed upon them. "Some of these species prey preferentially on large, mature-size classes of guppies." (I take all quotes from the primary technical article that inspired later press accounts—"Evaluation of the rate of evolution in natural populations of guppies *(Poecilia reticulata),*" by D. N. Reznick, F. H. Shaw, F. H. Rodd, and R. G. Shaw, published in *Science,* 1977, volume 275, pages

1934–37). Other populations of the same species live in "upstream portions of each drainage" where most "predators are excluded . . . by rapids or waterfalls, yielding low-predation communities."

In studying both kinds of populations, Reznick and colleagues found that "guppies from high-predation sites experience significantly higher mortality rates than those from low-predation sites." They then reared both kinds of guppies under uniform conditions in the laboratory, and found that fishes from high-predation sites in lower drainages matured earlier and at a smaller size. "They also devote more resources to each litter, produce more, smaller offspring per litter, and produce litters more frequently than guppies from low-predation localities."

This combination of observations from nature and the laboratory yields two important inferences. First, the differences make adaptive sense, for guppies subjected to greater predation would fare better if they could grow up fast and reproduce both copiously and quickly before the potential boom falls—a piscine equivalent of the old motto for electoral politics in Boston: vote early and vote often. On the other hand, guppies in little danger of being eaten might do better to bide their time and grow big and strong before engaging their fellows in any reproductive competition. Second, since these differences persist when both kinds of guppies are reared in identical laboratory environments, the distinction must record genetically based and inherited results of divergent evolution between the populations.

In 1981, Reznick had transferred some guppies from high-predation downstream pools into low-predation upstream waters then devoid of guppies. These transplanted populations evolved rapidly to adopt the reproductive strategy favored by indigenous populations in neighboring upstream environments: delayed sexual maturity at larger size, and longer life. Moreover, Reznick and colleagues made the interesting observation that males evolved considerably more rapidly in this favored direction. In one experiment, males reached their full extent of change within four years, while females continued to alter after eleven years. Since the laboratory populations had shown higher heritability for these traits in males than in females, these results make good sense. (Heritability may be roughly defined as the correlation between traits in parents and offspring due to genetic differences. The greater the heritable basis of a trait, the faster the feature can evolve by natural selection.)

This favorable set of circumstances—rapid evolution in a predictable and presumably adaptive direction based on traits known to be highly heritable—provides a "tight" case for well-documented (and sensible) evolution at scales well within the purview of human observation, a mere decade in this case. The headline for the news report on this paper in *Science* magazine (March 28, 1997) read: "Predator-free guppies take an evolutionary leap forward."

2. Lizards from the Exuma Cays, Bahama Islands. During most of my career, my fieldwork has centered on biology and paleontology of the land snail *Cerion* in the Bahama Islands. During these trips, I have often encountered fellow biologists devoted to other creatures. In one major program of research, Tom Schoener (a biology professor at the University of California, Davis) has, with numerous students and colleagues, been studying the biogeography and evolution of the ubiquitous little lizard *Anolis*—for me just a fleeting shadow running across a snail-studded ground, but for them a focus of utmost fascination (while my beloved snails, I assume, just blend into their immobile background).

In 1977 and 1981, Schoener and colleagues transplanted groups of five or ten lizards from Staniel Cay in the Exuma chain to fourteen small and neighboring islands that housed no lizards. In 1991, they found that the lizards had thrived (or at least survived and bred) on most of these islands, and they collected samples of adult males from each experimental island with an adequate population. In addition, they gathered a larger sample of males from areas on Staniel Cay that had served as the source for original transplantation in 1977 and 1981.

This study then benefits from general principles learned by extensive research on numerous *Anolis* species throughout the Bahama Islands. In particular, relatively longer limbs permit greater speed, a substantial advantage provided that preferred perching places can accommodate long-legged lizards. Trees and other "thick" perching places therefore favor the evolution of long legs. Staniel Cay itself includes a predominant forest, and the local *Anolis* tend to be long-legged. But when lizards must live on thin twigs in bushy vegetation, the agility provided by shorter legs (on such precarious perches) may outweigh the advantages in speed that longer legs would provide. Thus, lizards living on narrow twigs tend to be shorter-legged. The small cays that received the fourteen transported populations have little or no forest growth and tend instead to be covered with bushy vegetation (and narrow twigs).

J. B. Losos, the principal author of the new study, therefore based an obvious prediction on these generalities. The populations had been transferred from forests with wide perches to bushy islands covered with narrow twigs. "From the kind of vegetation on the new islands," Losos stated, "we predicted that the lizards would develop shorter hindlimbs." Their published study validates this expected result: a clearly measurable change, in the predicted and adaptive direction, in less than twenty years. (See details in J. B. Losos, K. I. Warheit, and T. W. Schoener, "Adaptive differentiation following experimental island colonization in *Anolis* lizards," *Nature,* 1997, volume 387, pages 70–73). A news report appeared in *Science* magazine (May 2, 1997) under the title "Catching lizards in the act of adapting."

This study lacks a crucial piece of documentation that the Trinidadian guppies provided—an absence immediately noted by friendly critics and fully acknowledged by the authors. Losos and colleagues have not studied the heritability of leg length in *Anolis sagrei* and therefore cannot be certain that their results record a genetic process of evolutionary change. The growth of these lizards may feature extensive flexibility in leg length, so that the same genes yield longer legs if lizards grow up on trees, and shorter legs if they always cavort in the bushes (just as the same genes can lead to a thin or fat human being depending upon a personal history of nutrition and exercise). In any case, however, a sensible and apparently adaptive change in average leg length has occurred within twenty years on several islands, whatever the cause of modification.

3. Snails from Great Inagua, Bahama Islands. Most of Great Inagua, the second-largest Bahamian Island (Andros wins first prize), houses a large and ribby *Cerion* species named *C. rubicundum.* But fossil deposits of no great age lack this species entirely and feature instead an extinct form named *Cerion excelsior,* the largest of all *Cerion* species. Several years ago, on a mudflat in the southeastern corner of Great Inagua, David Woodruff (of the University of California, San Diego) and I collected a remarkable series of shells that seemed to span (and quite smoothly) the entire range of form from extinct *C. excelsior* to modern *C. rubicundum.* Moreover, and in general, the more eroded and "older looking" the shell, the closer it seemed to lie to the anatomy of extinct *C. excelsior.*

This situation suggested a local evolutionary transition by hybridization, as *C. rubicundum,* arriving on the island from an outside source, interbred with indigenous *C. excelsior.* Then, as *C. excelsior* declined toward extinction, while *C. rubicundum* thrived and increased, the average anatomy of the population transformed slowly and steadily in the direction of the modern form. This hypothesis sounded good and sensible, but we could devise no way to test our idea—for all the shells had been collected from a single mudflat (analogous to a single bedding plane of a geological stratum), and we could not determine their relative ages. The pure *C. excelsior* shells "looked" older, but such personal impressions count for less than nothing (subject as they are to a researcher's bias) in science. So we got stymied and put the specimens in a drawer.

Several years later, I teamed up with paleontologist and geochemist Glenn A. Goodfriend from the Carnegie Institution of Washington. He had refined a dating technique based on changes in the composition of amino acids in the shell over time. By keying these amino acid changes to radiocarbon dates for some of the shells, we could estimate the age of each shell. A plot of shell age versus position on an anatomical spectrum from extinct *C. excelsior* to modern *C. rubicundum* produced a beautiful correlation between age and anatomy: the younger the specimen, the closer to the modern anatomy.

This ten- to twenty-thousand-year transition by hybridization exceeds the time period of the Trinidad and Exuma studies by three orders of magnitude (that is, by a factor of 1,000), but even ten thousand years represents a geological eye-blink in the fullness of evolutionary time—while this transformation in our snails marks a full change from one species to another, not just a small decrement of leg length, or a change in the timing of breeding, within a single species. (For details, see G. A. Goodfriend and S. J. Gould, "Paleontology and chronology of two evolutionary transitions by hybridization in the Bahamian land snail *Cerion*," *Science*, 1996, volume 274, pages 1894–97). Harvard University's press release (with no input from me) carried the headline "Snails caught in act of evolving."

A scanning of any year's technical literature in evolutionary biology would yield numerous and well-documented cases of such measurable, small-scale evolutionary change—thus disproving the urban legend that evolution must always be too slow to observe in the geological microsecond of a human lifetime. These three studies, all unusually complete in their documentation and in their resolution of details, do not really rank as "news" in the journalist's prime sense of novelty or deep surprise. Nonetheless, each of these three studies became subjects for front-page stories in either *The New York Times* or *The Boston Globe*.

Now please don't get me wrong. I do not belong to the cadre of rarefied academics who cringe at every journalistic story about science for fear that the work reported might become tainted with popularity thereby. And in a purely "political" sense, I certainly won't object if major newspapers choose to feature any result of my profession as a lead story—especially, if I may be self-serving for a moment, when one of the tales reports my own work! Nonetheless, this degree of public attention for workaday results in my field (however elegantly done) does fill me with wry amusement—if only for the general reason that most of us feel a tickle in the funny bone when we note a gross imbalance between public notoriety and the true novelty or importance of an event, as when Hollywood spinmeisters manage to depict their client's ninth marriage as the earth's first example of true love triumphant and permanent.

Of course I'm delighted that some ordinary, albeit particularly well done studies of small-scale evolution struck journalists as front-page news. But I still feel impelled to ask why these studies, rather than a hundred others of equal care and merit that appear in our literature every month, caught this journalistic fancy and inspired such prime attention. When I muse over this issue, I can only devise two reasons—both based on deep and interesting fallacies well worth identifying and discussing. In this sense, the miselevation of everyday good work to surprising novelty may teach us something important about public attitudes toward evolution, and toward science in general. We may, I think, resolve each of the two fallacies by contrasting the supposed meaning of these studies, as reported in public accounts, with the significance of such work as viewed by professionals in the field.

1. *The fallacy of the crucial experiment*

In high school physics classes, we all learned a heroically simplified version of scientific progress based upon a model that does work sometimes but by no means always—the *experimentum crucis,* or crucial experiment. Newton or Einstein? Ptolemy or Copernicus? Special Creation or Darwin? To find out, perform a single decisive experiment with a clearly measurable result replete with decisive power to decree yea or nay.

The decision to treat a limited and particular case as front-page news must be rooted in this fallacy. Reporters must imagine that evolution can be proved by a single crucial case, so that any of these stories may provide decisive confirmation of Darwin's truth—a matter of some importance given the urban legend that evolution, even if valid, must be invisible on human timescales.

But two counterarguments vitiate this premise. First, as a scientific or intellectual issue, we hardly need to "prove" evolution by discovering new and elegant cases. We do not, after all, expect to encounter a page-one story with the headline "New experiment proves earth goes around sun, not vice versa. Galileo vindicated." The fact of evolution has been equally well documented for more than a century.

Second, and more generally, single "crucial" experiments rarely decide major issues in science—especially in natural history, where nearly all theories require data about "relative frequencies" (or percentage of occurrences), not pristine single cases. Of course, for a person who believes that evolution never occurs at all, one good case can pack enormous punch—but science resolved this basic issue more than one hundred years ago. Nearly every interesting question in evolutionary theory asks "how often?" or "how dominant in setting the pattern of life?"—not "does this phenomenon occur at all?" For example, on the most important issue of all—the role of Darwin's own favored mechanism of natural selection—single examples of selection's efficacy advance the argument very little. We already know, by abundant documentation and rigorous theorizing, that natural selection can and does operate in nature. We need to determine the *relative strength* of Darwin's mechanism among a set of alternative modes for evolutionary change—and single cases, however elegant, cannot establish a relative frequency.

Professionals also commit this common error of confusing well-documented single instances with statements about relative strength among plausible alternatives. For example, we would like to know how often small and isolated populations evolve differences as adaptive responses to local environments (presumably by Darwin's mechanism of natural selection), and how often such changes occur by the random process known as "genetic drift"—a potentially important phenomenon in small populations (just as a small number of coin flips can depart radically from fifty-fifty for heads and tails, while a million flips with an honest coin cannot stray too far from this ideal). Losos's study on lizard legs provides one vote for selection (if the change turns out to have a genetic basis)—because leg length altered in a predicted direction toward better adaptation to local environments on new islands. But even such an elegant case cannot prove the domination of natural selection in general. Losos has only shown the power of Darwin's process in a particular example. Yet the reporter for *Science* magazine made this distressingly common error in concluding: "If it [change in leg length] is rooted in the genes, then the study is strong evidence that isolated populations diverge by natural selection, not genetic drift as some theorists have argued." Yes, strong evidence for these lizards on that island during those years—but not proof for the general domination of selection over drift. Single cases don't establish generalities, so long as alternative mechanisms retain their theoretical plausibility.

2. *The paradox of the visibly irrelevant*

As a second reason for overstating the centrality of such cases in our general understanding of evolution, many commentators (and research scientists as well) ally themselves too strongly with one of the oldest (and often fallacious) traditions of Western thought: reductionism, or the assumption that laws and mechanics of the smallest constituents must explain objects and events at all scales and times. Thus, if we can render the behavior of a large body (an organism or a plant, for example) as a consequence of atoms and molecules in motion, we feel that we have developed a "deeper" or "more basic" understanding than if our explanatory principles engage only large objects themselves, and not their constituent parts.

Reductionists assume that documenting evolution at the smallest scale of a few years and generations should provide a general model of explanation for events at all scales and times—so these cases should become a gold standard for the entire field, hence their status as front-page news. The authors of our two studies on decadal evolution certainly nurture such a hope. Reznick and colleagues end their publication on Trinidadian guppies by writing: "It is part of a growing body of evidence that the rate and patterns of change attainable through natural selection are sufficient to account for the patterns observed in the fossil record." Losos and colleagues say much the same for their lizards: "Macroevolution may just be microevolution writ large—and, consequently, insight into the former may result from study of the latter."

We tend to become beguiled by such warm and integrative feelings (for science rightly seeks unity and generality of explanation). But does integration by reduction of all scales to the rates and mechanisms of the smallest really work for evolution—and do we crave this style of unification as the goal of all science? I think not, and I also regard our best general reason for skepticism as conclusive for this particular subject—however rarely appreciated, though staring us in the face.

These shortest-term studies are elegant and important, but they cannot represent the general mode for building patterns in the history of life. The reason for their large-scale impotence strikes most people as deeply paradoxical, even quite funny—but the argument truly cannot be gainsaid. Evolutionary rates as measured for guppies and lizards are *vastly too rapid* to represent the general modes of change that build life's history through geological ages.

But how can I say such a thing? Isn't this statement ridiculous a priori? How could these tiny, minuscule changes—a little less leg, a minimally larger size—represent too much of anything? Doesn't the very beauty of these studies lie in their minimalism? We have always been taught that evolution is wondrously slow and cumulative—a grain-by-grain process, a penny a day toward the domain of Bill Gates. Doesn't each of these studies document a grain? Haven't my colleagues and I found the "atom" of evolutionary incrementation?

I believe that these studies have discerned something important, but they have discovered no general atom. These measured changes over years and decades are too fast by several orders of magnitude to build the history of life by simple cumulation. Reznick's guppy rates range from 3,700 to 45,000 darwins (a standard metric for evolution, expressed as change in units of standard deviation—a measure of variation around the mean value of a trait in a population—per million years). By contrast, rates for major trends in the fossil record generally range from 0.1 to 1.0 darwin. Reznick himself states that "the estimated rates [for guppies] are . . . four to seven orders of magnitude greater than those observed in the fossil record" (that is, ten thousand to ten million times faster).

Moreover and with complete generality—thus constituting the "paradox of the visibly irrelevant" in my title—we may say that any change measurable *at all* over the few years of an ordinary scientific study must be occurring far too rapidly to represent ordinary rates of evolution in the fossil record. The culprit of this paradox, as so often, can be identified as the vastness of time (a concept that we can appreciate "in our heads" but seem quite unable to place into the guts of our intuition). The key principle, however ironic, requires such a visceral understanding of earthly time: if a case of evolution proceeds with sufficient speed to be discerned by our instruments in just a few years—that is, if the change becomes substantial enough to stand out as a genuine and directional effect above the random fluctuations of nature's stable variation and our inevitable errors of measurement—then we have witnessed something far too substantial to serve as an atom of steady incrementation in a paleontological trend. Thus, to restate the paradox: if we can measure it at all (in a few years), it is too powerful to be the stuff of life's history.

If large-scale evolution proceeded by stacking Trinidad guppy rates end to end, then any evolutionary trend would be completed in a geological moment, not over the many million years actually observed. "Our face from fish to man," to cite the title of a famous old account of evolution for popular audiences, would run its course within a single geological formation, not over more than 400 million years, as our fossil record demonstrates.

Evolutionary theory must figure out how to slow down these measured rates of the moment, not how to stack them up! In fact, most lineages are stable (*non*changing) nearly all the time in the fossil record. When lineages do change, their alteration usually occurs "momentarily" in a geological sense (that is, confined to a single bedding plane) and usually leads to the origin of a new species by branching. Evolutionary rates during these moments may match the observed speed of Trinidadian guppies and Bahamian lizards—for most bedding planes represent several thousand years. But during most of a typical species's lifetime, no change accumulates, and we need to understand why. The sources of stasis have become as important for evolutionary theory as the causes of change.

(To illustrate how poorly we grasp this central point about time's immensity, the reporter for *Science* magazine called me when my *Cerion* article, coauthored with Glenn Goodfriend, appeared. He wanted to write an accompanying news story about the exception I had found to my own theory of punctuated equilibrium—an insensibly gradual change over ten to twenty thousand years. I told him that, although exceptions abound, this case does not lie among them, but actually represents a strong confirmation of punctuated equilibrium! We found all twenty thousand years' worth of snails on a single mudflat—that is, on what would become a single bedding plane in the geological record. Our *entire* transition occurred in a geological moment and represented a punctuation, not a gradual sequence of fossils. We were able to "dissect" the punctuation in this unusual case—hence the value of our publication—because we could determine ages for the individual shells. The reporter, to his credit, completely revised his originally intended theme and published an excellent account.)

In conclusion, I suspect that most cases like the Trinidadian guppies and Bahamian lizards represent transient and momentary blips and fillips that "flesh out" the rich history of lineages in stasis, not the atoms of substantial and steadily accumulated evolutionary trends. Stasis is a dynamic phenomenon. Small local populations and parts of lineages make short and temporary forays of transient adaptation, but these tiny units almost always die out or get reintegrated into the general pool of the species. (Losos himself regards the new island populations of lizards as evolutionarily transient in exactly this sense—for such tiny and temporary colonies are almost always extirpated by hurricanes in the long run. How, then, can such populations represent atoms of a major evolutionary trend? The news report in *Science* magazine ends by stating: "But whether the lizards continue to evolve depends largely on the winds of fate, says Losos. These islets are periodically swept by hurricanes that could whisk away every trace of anolian evolution.")

But transient blips and fillips are no less important than major trends in the total "scheme of things." Both represent evolution operating at a standard and appropriate measure for a particular scale and time—Trinidadian blips for the smallest and most local moment, faces from fish to human for the largest and most global frame. One scale doesn't translate into another. No single scale can be deemed more important than any other; and none operates as a basic model for all the others. Each scale embodies something precious and unique to teach us; none can be labeled superior or primary. (Guppies and lizards, in their exposition of momentary detail, give us insight, unobtainable at broader scales, into the actual mechanics of adaptation, natural selection, and genetic change.)

The common metaphor of the science of fractals—Mandelbrot's familiar argument that the coast of Maine has no absolute length, but depends upon the scale of measurement—epitomizes this principle well (see chapter 23). When we study guppies in a pond in Trinidad, we are operating at a scale equivalent to measuring the coastline by wrapping our string around every boulder on every headland of Acadia National Park. When we trace the increase in size of the human brain from Lucy (about four million years ago) to Lincoln, we are measuring the coastline as depicted on my page of Maine in *Hammond's Atlas.* Both scales are exactly right for their appropriate problems. You would be a fool to spend all summer measuring the details in one cove in Acadia, if you just wanted to know the distance from Portland to Machiasport for your weekend auto trip.

I find a particular intellectual beauty in such fractal models—for they invoke hierarchies of inclusion (the single cove embedded within Acadia, embedded within Maine) to deny hierarchies of worth, importance, merit, or meaning. You may ignore Maine while studying the sand grain, and be properly oblivious of the grain while perusing the map of Maine on the single page of your atlas. But you can love and learn from both scales at the same time. Evolution does not lie patent in a clear pond on Trinidad any more than the universe (*pace* Mr. Blake) lies revealed in a grain of sand. But how poor would be our understanding—how bland and restricted our sight—if we could not learn to appreciate the rococo details that fill our immediate field of vision, while forming, at another scale, only some irrelevant and invisible jigglings in the majesty of geological time.

23

Room of One's Own

GOLGOTHA, THE SITE OF CHRIST'S CRUCIFIXION, appears in most paintings as a substantial hill in the countryside, far from the city walls of Jerusalem depicted in a distant background. In fact, if the traditional spot has been correctly identified, Golgotha is a tiny protuberance located just next to the old city limits but now inside the walls built by Suleiman the Magnificent in the early sixteenth century. These walls extended the boundaries of Jerusalem, and the old town now sits as a small "jewel" at the center of a much bigger, modern city. Golgotha is small and low enough to fit *within* the Church of the Holy Sepulchre, located *within* Suleiman's city walls. Visitors just have to climb an internal staircase to reach the top of Golgotha, located on the church's second story. (Several theories compete to explain the derivation of the name, for *golgotha* means "skull" in Aramaic, while the alternative label of *calvary* has the same definition in Latin. Most scholars think that the name designates the shape of the small hill, not the mortal remains of executions.)

As one of the most sacred sites on earth, the Church of the Holy Sepulchre might be expected to exude dignity, serenity, and a spirit of transcendence above merely earthly cares. Yet in maximal, almost perverse contrast, the church is a site of constant bickering and division. The etymology of *religion* may refer to "tying together," but the actual experience, given the propensities of *Homo sapiens,* the earth's most various and curmudgeonly species, tends more often to separation and anathematization. The precious space is "shared" (in this case, a euphemism for "wrangled over") by six old Christian groups—Greek Orthodox, Roman Catholic, Armenian, Syrian, Coptic, and Abyssinian. (The various Protestant denominations came upon the scene a few centuries too late and didn't even get a pew.)

Before visiting the church several years ago, I had encountered the Latin phrase *status quo* only as a general description for leaving well enough alone. But I learned that the phrase can also be used as a proper noun—capital *S,* capital *Q.* In 1852, after centuries of more serious bickering, the six groups signed an agreement, called *the* Status Quo, to regulate every move and every square inch in the building. At this point, I will yield to Baedeker's *Guide to Jerusalem* (1982 edition), a publication generally known for authoritative and stodgy prose but uncharacteristically pungent in this case:

> No lamp, no picture, nothing whatsoever may be moved without its giving rise to a complaint. The rules governing when and where each community may celebrate Mass are minutely prescribed as are the times when the lamps may be lit and the windows may be opened. Everything must be done in accordance with the originally agreed rules, i.e. the "status quo." . . . Modifications to this are persistently being sought and just as persistently rejected—they even cropped up in the negotiations for the Treaty of Versailles and in the League of Nations. . . . Anyone hoping to find harmony and quiet contemplation . . . is due for a disappointment—the sects are on a Cold War footing. Even the background noise can be put down to psychological warfare—the sound of the blows of hammers and chisels constantly engaged on improvement work mingles with the chanting of Greek plainsong, blasts from the Franciscan organ and the continual tinkling of Armenian bells.

And lest anyone hope that equality might reign among the six groups, I hasten to point out that the Status Quo assigned 65 percent of the church to the

Greek Orthodox, while granting the Abyssinians—the only black African group by ethnicity—just the tomb of Joseph of Arimathea ("a tiny cavity that can only be reached by passing through Coptic territory," to quote Baedeker's one more time). Adding insult to this injury, the poor Abyssinians can't even reside within the church but must live instead in tiny cells built on the roof! (And let me tell you, it was really hot up there the day I visited.)

To move from a ridiculous story about a sublime place to the fully ridiculous all around, I got the idea for this essay from an English newspaper story of July 9, 1997: "Punch-up Between Brewery Rivals Over Future of Historic Hostelry." One of London's most interesting pubs, the Punch Tavern on Fleet Street, bears a name that reflects a former role as the favorite watering hole for staff members of the famous humor magazine. These ghosts of the past could have filed quite a story on the current situation. Bass, a large national brewery, owns two-thirds of the property, including the only toilets. But Samuel Smith, a smaller, regional operation, bought the other third, including the passageway for delivery of beer to the Bass side. The two businesses have coexisted in constant tension and bickering but have now opted for something closer to the Holy Sepulchre solution of strict division. A new wall now rises within the pub, and the Bass people are building "a new cellar drop so workers can move beer supplies without using Samuel Smith's passageway." We must assume that the Smith folks will construct some new toilets, for we all know that such items rank second only to what comes in the other end as a necessary fixture in these establishments.

One last item, ridiculous but personal this time, will serve to establish this theme as a generality. My brother and I shared a small room throughout our childhood. We usually coexisted reasonably well, but we did have our battles from time to time. One day, following our worst blowup, Peter decided that we would just have to divide the room right down the middle, and each of us promise never to set so much as a toe into the other's territory. He proceeded to gather all his possessions and move them to his side. But I just lay on my bed laughing—as he got progressively angrier at my lighthearted approach to such a serious situation. When he finished all the moving and shoving, he confronted me in a fury: "What are you laughing about?" I didn't say a word, but only lifted my finger and pointed at the room's single door—located on my side. Fortunately, Peter started to laugh too; so we made up and amalgamated all our stuff again.

If people, representing a mere few billion souls within a single species spread throughout the planet, can generate so much strife about divvying a space, what can nature possibly do with millions of species, gazillions of individuals, and nothing with the ability or power to negotiate, or even to understand, a status quo? Much of ecological theory has been devoted to debating concepts that may usually be framed differently, but really represent variants of this fundamental question.

Consider just two examples that generally make the rounds in any basic college course on evolution. In discussing the crucial question of how many species can coexist in a single habitat (obviously an ever more important issue as natural spaces shrink before human onslaught, and many species face imminent extinction), students invariably hear about something called the "competitive exclusion" principle, or the notion that two species cannot occupy the same "niche." This conclusion follows more as a logical consequence of natural selection than an observation from nature. If two species lived in truly identical environments, sharing all the same spaces and resources, then one of the two would surely hold some advantage, however slight, over the other, and the relentless force of natural selection, acting on even the tiniest differential over countless generations, should secure total victory for the species with a small edge in the competitive struggle for existence.

But this principle probably says less than its weighty words seem to imply, for niches do not exist independently of the species that inhabit them. Niches are not comparable to houses in a suburban development, built "on spec" and fully decked out with all furnishings and utilities before people come to buy under a strict rule of "one lot for one family." Niches are constructed by organisms as they interact with complex environments—and how could two different species read an environment in exactly the same way for all particulars?

A related principle (and second example) called "limiting similarity" attempts to put this theme into a more reasonable and testable light. If two separate species cannot be identical in appearance and behavior, and cannot read the surrounding environment in exactly the same way, then how close can they be? What are the limits to their similarity? How many species of beetles can live in a tropical tree? How many species of fishes in a temperate pond?

We can at least pose such a question without logical contradiction, and we can test certain ideas about minimal discrepancies in body size, feeding preferences, and so on. Much useful research has been done on this subject, but no general answers have emerged. And none may be possible (at least in such simplistic form as "no more similar than a 10 percent difference in body weight on average"), given the irreducible uniqueness of each species and each group of organisms. Beetle rules will almost surely not work as fish rules, not to mention the vastly more different world of rules for bacteria.

But if we cannot generate quantitative laws of nature about numbers of species in a single place, we can at least state some general principles. And the rule behind Jerusalem's Status Quo, whatever its moral dubiety in the ethical systems of *Homo sapiens,* provides a good beginning: large numbers of species can be crammed into a common territory only if each can commandeer some room of its own and not always stand in relentless competition with a maximally similar form.

Two general strategies may be cited, the second far more interesting than the first, for acquiring the requisite "breathing room"—a little bit of unique space that no other species contests in exactly the same way. In the first strategy—the "Holy Sepulchre solution" if you will—two species perceive the surrounding environment in basically the same manner and therefore must divide the territory to keep out of each other's way. Division may be strictly spatial, as in my fraternal dispute about our single common room. But organisms may also use nature's other prime dimension and construct temporal separations as well. The Status Quo divides the space within the Church of the Holy Sepulchre, but the agreement also decrees when the unitary domain of sound belongs to the masses, instruments, and voices of various competing groups.

To make an ugly analogy, based on cruel social practices now thankfully abandoned, but in force not long ago, I encountered both spatial and temporal modes of segregation when I began my college studies in southwestern Ohio during the late 1950s. The town movie theater placed whites in the orchestra and blacks in the balcony, while the local skating rinks and bowling alleys maintained different "white" and "Negro" nights. (Student and community activism, spurred by the nascent civil rights movement, fought and vanquished these cruelties during my watch. I remember my own wholehearted and, in retrospect, pretty inconsequential participation with pride.)

An instructive evolutionary example of this first strategy arises from a classical argument about modes of speciation, or the origin of a new species by branching from an ancestral population. Such branching may occur if a group of organisms can become isolated from the parental population and begin to breed only among themselves in a different environment that might favor the evolution of new features by natural selection. (If members of the separating group continue to interact and breed with individuals of the parental population, then favorable new features will be lost by dilution and diffusion, and the two groups will probably reamalgamate, thus precluding the origin of a new species by branching.)

The conventional theory for speciation—called allopatric, and meaning "living in another place"—holds that a population can gain the potential to form a new species only by becoming geographically isolated from the ancestral group, for only strict spatial separation can guarantee the necessary cutoff from contact with members of the parental population. Much research into the process of speciation has focused on the modes of attaining such geographic isolation—new islands rising in the sea, continents splitting, rivers changing their courses, and so on.

A contrasting idea—called sympatric speciation, or "living in the same place"—holds that new groups may speciate while continuing to inhabit the same geographic domain as the parental population. The defense of sympatric speciation faces a classic conundrum, and most research on the subject has been dedicated to finding solutions: if isolation from members of the parental population is so crucial to the formation of a new species, how can a new species arise within the geographic range of the parents?

This old issue in evolutionary theory remains far from resolution, but we should note, in the context of this essay, that proposed mechanisms usually follow the Holy Sepulchre principle of granting the new group a room of its own within the spatial boundaries of the parental realm—and that such "internal isolation" may be achieved by either the spatial or the temporal route. The best-documented cases of the spatial strategy invoke a process with the technical name of *host specificity,* or the restriction of a population to a highly specific site within a general area. For example, to cite an actual (although still controversial) case, flies of the genus *Rhagoletis* tend to inhabit only one species of tree as an exclusive site for breeding and feeding. Suppose that some individuals within a species that lives on apple trees experience a mutation that leads them to favor hawthorns. A new population, tied exclusively to hawthorns, may soon arise and may evolve into a separate species. The hawthorn flies live within the same geographic region as the apple flies, but members of the two groups never interbreed because each recognizes only a portion of the total area as a permissible home—just as the six sects of the Holy Sepulchre never transgress into one another's territory.

The same principle may work temporally as well. Suppose that two closely related species of frogs live and reproduce in and around the same pond, but one species uses the day-lengthening cues of spring to initiate breeding, while the other waits for the day-shortening signals of fall. The two populations share the same space and may even (metaphorically) wave and wink at each other throughout the year, but they can never interbreed and can therefore remain separate as species.

In the second, and philosophically far more interesting, strategy for securing a requisite room of one's own, species may share the same region but avoid the need for a natural equivalent of the Status Quo, because they do not perceive each other at all and therefore cannot interfere or compete—blessedly benign ignorance rather than artfully negotiated separation. This fascinating form of imperception, which can also be achieved by either spatial or temporal routes, raises one of the most illuminating issues of intellectual life and nature's construction: the theme of scaling, or strikingly different ways of viewing the world from disparate vantage points of an observer's size or life span, with no single way either universally "normal" or "better" than any other.

To begin with a personal story, I share my Harvard office with about a hundred thousand trilobites, all fossils at least 250 million years old, and now housed in cabinets lining the perimeter of my space. For the most part, we coexist in perfect harmony. They care little for my eye-blink of a forty-year career, and I view them with love and respect to be sure, but also as impassive, immobile pieces of rock. They cause me no trouble because I just move the appropriate drawers to an adjacent room when a visiting paleontologist needs to study a genus or two. But one week, about ten years ago, two British visitors wanted to look at *all* Ordovician trilobites, an endeavor that required exploratory access to all drawers. I had no choice but to abandon my office for several days—a situation made worse by the stereotypical politeness of my visitors, as they apologized almost hourly: "Oh, I do hope we're not disturbing you too much." I wanted to reply: "You bloody well are, but there's nothing I can do about it," but I just shut up instead. I relaxed when I finally figured out the larger context. My visitors, of course, had been purposely sent by the trilobites to teach me the following lesson of scaling: we will let you borrow this office for a millimoment of our existence; this situation troubles us not at all, but we do need to remind you about the room's true ownership once every decade or so, just to keep you honest.

Species can also share an environment without conflict when each experiences life on such a different temporal scale that no competitive interaction ever occurs. A bacterial life cycle of half an hour will pass beneath my notice and understanding, unless the population grows big enough to poison or crowd out something of importance to me. And how can a fruit fly ever experience me as a growing, changing organism if I manifest such stability throughout the fly's full life cycle of two weeks or so? The pre-Darwinian Scottish evolutionist Robert Chambers devoted a striking metaphor to this point when he wondered if the adult mayfly, during its single day of earthly life, might mistake the active metamorphosis of a tadpole into a frog for proof of the immutability of species, since no visible change would occur during the mayfly's entire lifetime. (And so, Chambers argued by extension, we might miss the truth of evolution if the process unrolled so slowly that we could never notice any changes during the entire history of potential human observation.) Chambers wrote in 1844:

> Suppose that an ephemeron [a mayfly], hovering over a pool for its one April day of life, were capable of observing the fry of the frog in the waters below. In its aged afternoon, having seen no change upon them for such a long time, it would be little qualified to conceive that the external branchiae [gills] of these creatures were to decay, and be replaced by internal lungs, that feet were to be developed, the tail erased, and the animal then to become a denizen of the land.

Since organisms span such a wide range of size, from the invisible bacterium to the giant blue whale (or to the fungus that underlies a good part of Michigan), the second, or spatial, strategy of coexistence by imperception achieves special prominence in nature. This concept can best be illustrated by an example that has become something of a cliché (by repetition for appropriateness) in intellectual life during the past decade.

To illustrate his concept of "fractals," mathematical curves that repeat an identical configuration at successively larger or smaller scales ad infinitum, mathematician Benoit Mandelbrot asked a disarmingly simple question with a wonderfully subtle nonanswer: how long is the coastline of Maine? The inquiry sounds simple but cannot be resolved without ambiguity, for solutions depend upon the scale of inquiry, and no scale can claim a preferred status. (In this respect, the question recalls the classic anecdote, also told about folks "down East" in Maine, of a woman who asks her neighbor, "How's your husband?"—and receives the answer, "Compared to what?")

If I'm holding an atlas with a page devoted to the entire state of Maine, then I may measure a coastline at the level of resolution permitted by my source. But if I use a map showing every headland in Acadia National Park, then the equally correct coastline becomes much longer. And if I try to measure the distance around every boulder in every cove of Acadia, then the length becomes ever greater (and increasingly less meaningful as tides roll and boulders move). Maine has no single correct coastline; any proper answer depends upon the scale of inquiry.

Similarly for organisms. Humans rank among the largest animals on earth, and we view our space as one might see all of Maine on a single page. A tiny organism, living in a world entirely circumscribed by a single boulder in a cove, will therefore be completely invisible at our scale. But neither of us sees "the world" any better or any more clearly. The atlas defines my appropriate world, while the boulder defines the space of the diatom or rotifer (while the rotifer then builds the complete universe of any bacterium dwelling within).

We need no Status Quo to share space with a bacterium, for we dwell in different worlds of a common territory—that is, unless we interfere or devise a way to intrude: the bacterium by generating a population large enough to incite our notice or cause us harm; *Homo sapiens* by inventing a microscope to penetrate the world of the invisible headland on a one-page map of the earth.

Frankly, given our aesthetic propensities, we would not always wish to perceive these smaller worlds within our domain. About 40 percent of humans house eyebrow mites, living beneath our notice at the base of hair follicles above our eyes. By ordinary human standards, and magnified to human size, these mites are outstandingly ugly and fearsome. I would just as soon let them go their way in peace, so long as they continue the favor of utter imperceptibility. And do we really want to know the details of ferocious battles between our antibodies and bacterial invaders—a process already distasteful enough to us in the macroscopic consequence of pus? (Don't get me wrong. As a dedicated scientist, I do assert the cardinal principle that we always want to know intellectually, both to understand the world better and to protect ourselves. I am just not sure that we should always crave visceral perception of phenomena that don't operate at our scale in any case.)

Finally, this theme of mutually invisible life at widely differing scales bears an important implication for the "culture wars" that supposedly now envelop our universities and our intellectual discourse in general (but that have, in my opinion, been grossly oversimplified and exaggerated for their perceived newsworthiness). One side of this false dichotomy features the postmodern relativists who argue that all culturally bound modes of perception must be equally valid, and that no factual truth therefore exists. The other side includes the benighted, old-fashioned realists who insist that flies truly have two wings, and that Shakespeare really did mean what he thought he was saying. The principle of scaling provides a resolution for the false parts of this silly dichotomy. Facts are facts and cannot be denied by any rational being. (Often, facts are also not at all easy to determine or specify—but this question raises different issues for another time.) Facts, however, may also be highly scale dependent—and the perceptions of one world may have no validity or expression in the domain of another. The one-page map of Maine cannot recognize the separate boulders of Acadia, but both provide equally valid representations of a factual coastline.

Why should we privilege one scale over another, especially when a fractal world can express the same form at every scale? Is my hair follicle, to an eyebrow mite, any less of a universe than our entire earth to the Lord of Hosts (who might be a local god as tiny as a mite to the great god of the whole universe—who then means absolutely nothing in return to the mite on my eyebrow)? And yet each denizen of each scale may perceive an appropriate universe with impeccable, but local, factual accuracy.

We don't have to love or even to know about all creatures of other scales (although we have ever so much to learn by stretching our minds to encompass, however dimly and through our own dark glasses, their equally valid universes). But it is good and pleasant for brethren to dwell together in unity—each with some room of one's own.

ILLUSTRATION CREDITS

Grateful acknowledgment is made to the following for permission to reproduce the images herein:

pages v, 14	American Museum of Natural History, photograph by Jackie Beckett
page 37	American Museum of Natural History, photograph by Stephanie Bishop
page 38	The Granger Collection, New York
page 96	Rare Book Collection, Skillman Library, Lafayette College
page 127	Courtesy of Jonathan A. Hill
page 128	Christie's Images
page 193	American Museum of Natural History, photographs by Jackie Beckett
pages 204, 208, 212, 213, 214	All images from the Edward Arnold Collection, courtesy of the Eiffel Tower Millennial Exhibition
page 300	Corbis-Bettmann
page 309	Corbis-Bettmann
page 322	Courtesy of Joyce Pendola
pages 323, 330 (bottom images)	Courtesy of Andrew Knoll

All other images appearing throughout are from the author's collection.

INDEX

C

H

I

M

N

O

P

T